MEDICINAL PLANTS OF ICELAND

Arnbjörg Linda Jóhannsdóttir

MEDICINAL PLANTS OF ICELAND

Collection, preparation and uses

MÁL OG MENNING

Medicinal Plants of Iceland
Collection, preparation and uses

Mál og menning
Reykjavík 2012

Book and Cover Design: Alexandra Buhl / Forlagið
Body text font: CamingoDos Pro Cd, 9pt
Layout: Einar Samúelsson / Hugsa sér!
Illustrations: Shutterstock, Wikimedia, except page 14 © Guðmundur Hafsteinsson,
page 52 and page 70 © Anna Birna Jóhannesdóttir.
Printing: Almarose, Slovenia

ISBN 978-9979-3-3278-7

Mál og menning is an imprint of Forlagið ehf.
www.forlagid.is

CONTENTS

FOREWORD

The use of herbs to cure is almost as old as mankind itself, and mention is made widely of such in ancient texts. Humans are thought to have originally gathered and eaten many plants as food, gradually learning to utilise them to heal wounds and treat various ailments. They discovered which plants were good to eat, which had medicinal properties and which were poisonous. Through the ages, this knowledge was passed on, and added to, by successive generations.

The oldest sources on herbal medicine are around 5000 years old, and come from China, Babylon, Egypt and India. Much of the knowledge acquired by these peoples of medicinal herbs has, as might be expected, been lost through time. Today we look primarily to the Chinese, the French and native Americans.

When herbs are used for medicinal purposes, the prime emphasis is on using certain parts in their entirety, rather than isolated active substances which have the strongest effect. The variety of substances contained in the herbs have a wide-reaching beneficial effect on the entire body, as is most clearly evident when using, for example, garlic, dandelion and yarrow. All of these herbs can be used for many types of illnesses and physical disorders.

In Iceland, the tradition of preparing and using herbal medicines is as old as the settlement of the country itself and is mentioned in both medieval sagas and folktales through the centuries. Icelandic knowledge of herbs and their uses originally came from the Nordic countries and British Isles, but later from more southerly reaches. According to written sources, Hrafn Sveinbjarnarson, one of the most renowned physicians of the Commonwealth period in Iceland (from the 9th to the 13th century), obtained his knowledge from the medical school at Salerno, Italy. There are no reports of Icelandic physicians growing herbs specifically for their own use, but they are known to have made considerable use of those growing in the wild.

Late in the 18th century, Rev. Björn Halldórsson, an early agricultural pioneer in Iceland, published *Grasnytjar*, or the use every farmer can have of those repetiton of those wild plants growing on his lands. His book explained how herbs could be gathered and used as food, medicine, dye and even to repel pests.

It is my sincere hope that this book will increase interest and knowledge of herbal medicines among the general public, as there is no denying the beneficial effect of many herbs, as long as they are properly used and understood.

Arnbjörg Linda Jóhannsdóttir

ARNBJÖRG LINDA JÓHANNSDÓTTIR

Arnbjörg Linda Jóhannsdóttir was born in Bíldudalur, in West Iceland, in 1959. Upon concluding studies at the School of Herbal Medicine in Kent, England, 1984-1987, she immediately began work as an herbalist in Iceland, which she has pursued ever since. In 1990-1994 she returned to England to study traditional Chinese medicine and acupuncture at the International College of Oriental Medicine. This has enabled her to combine the knowledge and techniques of both herbal medicine and acupuncture in her work since then. Arnbjörg Linda Jóhannsdóttir is the mother of three daughters.

ICELANDIC HERBS

Bistorta vivipara – Buckwheat family (Polygonaceae)

ALPINE BISTORT

OSTERICK, SNAKEWEED, ADDERWORT

RANGE AND HABITAT: Very common throughout Iceland. Grows in all types of soils, both in the highlands and settled areas.

PARTS USED MEDICINALLY: Root.

GATHERING: Early summer.

ACTIVE SUBSTANCES: Tannins, starch and gallic acid.

MEDICINAL ACTION: Strongly astringent and demulcent.

USES: The root is used for all types of pain and haemorrhage in the digestive tract, e.g. for gastritis and colitis, and for diarrhoea.

The root is also useful for gingivitis and other soreness in the mouth and throat, as well as for vaginal and cervical soreness. A decoction of the root can be useful to treat sores which are hard to heal.

DOSAGES:

TINCTURE: 1:5, 25% alcohol, 1-2 ml three times daily.
DECOCTION: 1:10, 20-40 ml three times daily or 1/2 tsp in 1 cup of water, drunk three times daily.
Crushed root for external use.
Infusion and diluted tincture as a rinse.

- Smaller doses are required for children.
- The three-sided seeds are nutritious and were formerly eaten either on their own or in milk.

Diphazium alpinum – (Lycopodiaceae

ALPINE CLUBMOSS

RANGE AND HABITAT: Grows in hollows and depressions in the snowier parts of Iceland.

PARTS USED MEDICINALLY: Spores.

GATHERING: Late summer.

ACTIVE SUBSTANCES: Oils containing, for instance, palmitin glycerol, stearin, arachitine substances and lycopodium oil, also lycopodic acid and pollenin. In addition, the entire plant contains alkaloids, including clavatoxin and clavatin.

MEDICINAL ACTION: Vulnerary, stops bleeding and laxative, mildly antispasmodic.

USES: The spores were used to treat nephritis, cystitis and hepatitis. It was also recommended for incontinence.

Today, alpine clubmoss spores are used almost exclusively externally, to heal bad wounds and skin ailments, especially eczema and psoriasis.

DOSAGES:

Dried spores for external use.

- The plant is not recommended for children.
- Related types of clubmosses, such as wolf's foot clubmoss (*Lycopodium clavatum*) and stiff clubmoss (*L. annotinum*) have the same characteristics as alpine clubmoss. Both of these species are much rarer than alpine club-moss, in addition to which wolf's foot clubmoss is completely protected in Iceland.
- As the Icelandic name "dyeing clubmoss" indicates, the plant was used as a dye, most often with other colour sources.

WARNING!

The plant itself is considerably toxic, although the spores are not. The spores can only be used externally and in small doses.

Erigeron borealis – Daisy family (Asteraceae)

ALPINE FLEABANE

RANGE AND HABITAT: Common throughout Iceland. Grows on slopes and grassy moors.

PARTS USED MEDICINALLY: The whole plant, with the exception of the root.

GATHERING: Early summer.

ACTIVE SUBSTANCES: Volatile oils, rutin, alkaloids, mucilage and various minerals.

MEDICINAL ACTION: Astringent, expectorant, stimulates menstruation, laxative, skin demulcent.

USES: Alpine fleabane can be dangerous if used internally. It was used in this way formerly, but it is now evident that protracted use can cause hepatic toxicity, which could even be fatal.

Used externally, alpine fleabane is very useful to treat burns, inflammation and melanoma ulcers. Poultices of the herb are also useful in treating rheumatic inflammation and sore muscles. Alpine fleabane is useful as a gargle for mouth and throat ulcers and inflammation and as a vaginal rinse.

DOSAGES:

INFUSION for a rinse: 1:15 boiled water.
Poultices for external use.

- The plant is not recommended for children.
- The Icelandic name means "James's dandelion" and probably refers to the apostle James.

WARNING!
Oral consumption of alpine fleabane is dangerous due to its toxic effect on the liver.

Alchemilla alpina – Rose family (Rosaceae)

ALPINE LADY'S MANTLE

RANGE AND HABITAT: Grows in many types of dry soil. Common everywhere in Iceland.

PARTS USED MEDICINALLY: The leaves just before flowering.

GATHERING: Early summer.

ACTIVE SUBSTANCES: Tannins and unknown anti-inflammatory agents.

MEDICINAL ACTION: Vulnerary, anti-inflammatory, astringent and constrictive externally and internally, strengthens all body tissue.

USES: Alpine lady's mantle was formerly used extensively for diarrhoea and for pain and bleeding in the digestive tract. Externally lady's mantle is used to stop bleeding and flow of pus from wounds, as well as to heal minor sores. Both powder and poultices can be used.

The plant can be used as a gargle for a sore and inflamed mouth and as a rinse to remedy vaginal pain.

CHINESE MEDICINE: Alpine Lady's mantle strengthens the Spleen.

DOSAGES:

TINCTURE: 1:5, 25% alcohol, 2-4 ml three times daily.
INFUSION: 1:10, 50-75 ml three times daily or 1/2-1 tsp in 1 cup of water, drunk three times daily.
Poultices and powder for external use.
Infusion and diluted tincture as a rinse for external use.

- Smaller doses are required for children.
- No attempt should be made to staunch haemorrhaging in the digestive tract or diarrhoea without seeking the advice of a physician for the ailment.
- The herb was formerly considered effective against all types of throat ailments and was also called "throat herb".

Phleum alpinum – Grass family (Poaceae)

ALPINE TIMOTHY

RANGE AND HABITAT: Grows primarily in fertile mountain meadows. Common throughout Iceland.

PARTS USED MEDICINALLY: Leaves.

GATHERING: Early summer.

ACTIVE SUBSTANCES: Not researched.

MEDICINAL ACTION/USAGE: If the plant is gathered before the shoot tip flowers it can be useful as an expectorant. It is also a diuretic and sudorific and relieves flatulence. The herb has been used against colds, especially bronchitis and constipation with good success. The plant is also considered useful for skin irritation, both internally and externally.

CHINESE MEDICINE: Alpine timothy is warming and corrects Liver energy.

DOSAGES:

TINCTURE: 1:5, 25% alcohol, 1-3 ml three times daily.
INFUSION: ½ tsp: cup of water, drunk three times daily.
Liniment for external use.

- Smaller doses are required for children.

Papaver radicatum – Poppy family (Papaveraceae)

ARCTIC POPPY

ROOTED POPPY, YELLOW POPPY

RANGE AND HABITAT: Grows in gravelly areas and scree. Common in the West Fjords and in the south of East Iceland, rarer in other parts of the country.

PARTS USED MEDICINALLY: Flowers.

GATHERING: Late summer.

ACTIVE SUBSTANCES: Vegetable alkalis, tannins and resins.

MEDICINAL ACTION/USAGE: Arctic poppy is analgesic, sedative and astringent.

The flowers were used for all sorts of pains and were also recommended for internal haemorrhaging, especially in the digestive tract.

The flowers were also considered useful for insomniacs.

CHINESE MEDICINE: Arctic poppy is acidic, astringent and neutral. It affects the Kidneys, Intestines and Lungs.

DOSAGES:

TINCTURE: 1:5, 45% alcohol, 1-2 ml three times daily.

INFUSION: 1:10, 20-30 ml twice to three times daily or 1/2 tsp in 1 cup of water, drunk twice to three times daily.

- The plant is not recommended for children.
- Arctic poppy is thought to contain soporific and anaesthetic substances; it was formerly also called "sleeping herb".

Pyrola grandiflora – Wintergreen family (Pyrolaceae)

ARCTIC WINTERGREEN

RANGE AND HABITAT: Widespread in North and Northeast Iceland, rare elsewhere. Grows in brush and moist moorlands.

PARTS USED MEDICINALLY: Leaves.

GATHERING: In the summer prior to flowering.

ACTIVE SUBSTANCES: The plant has not been researched extensively, but is known to contain tannins.

MEDICINAL ACTION: Astringent, roborant, diuretic and anti-spasmodic.

USES: Wintergreen is considered effective to relieve pain and inflammation of the digestive tract. It has also been used for pain in the urinary tract, especially pain resulting from kidney stones or sand. Wintergreen is thought to have a strengthening effect on the womb, and has therefore been used to counteract women's heavy menstruation or breakthrough bleeding. Wintergreen is effective on wounds, and a tea or ointment of the herb is used for many types of eczema or sores which are slow to heal.

A weak infusion of the leaves is considered a good wash for sensitive or swollen eyes.

CHINESE MEDICINE: Wintergreen is cooling and strengthens the Liver.

DOSAGES:

TINCTURE: 1:5, 25% alcohol, 2-3 ml three times daily.
INFUSION: 1:10, 25-50 ml three times daily or 1/2-1 tsp in 1 cup of water, drunk three times daily.
Tea, poultices and ointment for external use.
Infusion and diluted tincture as eyewash.

- Smaller doses are required for children.

Arctostaphylos uva-ursi – Heather family (Ericaceae)

BEARBERRY

RANGE AND HABITAT: Heath moorlands and woodlands. Common almost everywhere in Iceland.

PARTS USED MEDICINALLY: Fresh or dried leaves.

GATHERING: Early summer.

ACTIVE SUBSTANCES: Arbutin, methyl arbutin, flavonoids, tannins, resins, allantoin, volatile oils and various organic acids.

MEDICINAL ACTION Cleanses the urinary tract of bacteria as well as being diuretic, astringent and demulcent.

USES: Bearberry is used primarily to eliminate urinary infections and cystitis. The body converts the arbutin to hydroquinone, which is antiseptic. Hydroquinone works best in an acidic environment.

Bearberry can also be used for kidney stones and sand in the urinary organs, it softens them and facilitates their passage through the urinary tract.

DOSAGES:

TINCTURE: 1:5, 45% alcohol, 1-3 ml three times daily.
INFUSION: 1:10, 50 ml three times daily or 1 tsp in 1 cup of water, drunk three times daily.

- The plant is not recommended for children.
- In earlier times the leaves of bearberry were used extensively for dyeing and making ink and the berries for tanning.

WARNING!
Never consume bearberry for a continuous lengthy period, as this can cause toxic effects due to hydroquinone.

BILBERRY

BLUE WHORTLEBERRY

RANGE AND HABITAT: Common in thickets and brush, in bogs and moorlands, where protected by snow covering in winter.

PARTS USED MEDICINALLY: Leaves and berries.

GATHERING: Early August.

ACTIVE SUBSTANCES: Organic acids, mucilage, sugars, various minerals, tannins and vitamins A, B and C.

MEDICINAL ACTION: Astringent, antiseptic, cooling. The leaves lower blood sugar levels.

USES: Dried berries and leaves are a very effective remedy for diarrhoea and also intestinal pain and colitis.

They can be used as a mouthwash for sore and swollen gums. Fresh berries appear to promote regular bowel movements (although if consumed in great quantities they can cause diarrhoea in some people). Fresh berries stimulate appetite and are thought to eliminate ascarids.

The leaves are useful to remedy cystitis (especially if caused by *E. coli* bacteria), prostatitis and lower blood sugar levels.

CHINESE MEDICINE: The leaves are cooling and can be used to counteract Heat in the Bladder, Colon and Prostate.

DOSAGES:

TINCTURE of dried berries and leaves: 1:5, 45% alcohol, 2-3 ml three times daily.
INFUSION: 1:10, 25-50 ml three times daily or 1 tsp in 1 cup of water, drunk three times daily. Infusion and diluted tincture as mouthwash. Fresh berries are eaten.

- The berries are also good in jams, soups and wine.

WARNING!
The leaves can cause toxicity/poisoning symptoms if used for more than 3-4 weeks at a time. Leaves are not recommended for children.

Empetrum nigrum – Heather family (Ericaceae)

BLACK CROWBERRY

RANGE AND HABITAT: Grows in many types of soil throughout Iceland.

PARTS USED MEDICINALLY: Berries.

GATHERING: Autumn.

ACTIVE SUBSTANCES: There is little research on the chemical content of the berries, but they are known to contain both tannins and vitamins.

MEDICINAL ACTION/USAGE: Juice of the berries is very astringent and therefore useful to treat pain and haemorrhaging in the digestive tract. It is also good to regulate bowel movements, especially in children. The berries are rich in vitamins and therefore nutritious if eaten in moderation.

Excessive consumption can be overly constrictive and result in constipation and lack of appetite.

DOSAGES:

2 tbsp of berry juice every other hour for diarrhoea.

- Crowberries were formerly used as a dye and are considered good for winemaking.

Fucus vesiculosus – Algae family (Fucaceae)

BLADDERWRACK

BLACK TANG

RANGE AND HABITAT: Grows on rocky beaches, common throughout Iceland on such coastline.

PARTS USED MEDICINALLY: Entire plant. Kelp tablets are available in all health food stores and some pharmacies.

GATHERING: Year-round.

ACTIVE SUBSTANCES: Mucilage, mannitol, fucosterin, fucoxanthin, sisantin, volatile oils and many minerals, e.g. iodine.

MEDICINAL ACTION: Nourishing, purges the blood, stimulates the thyroid gland and thereby metabolism.

USES: Bladderwrack is used to treat an underactive thyroid gland (goitre). It is also useful for chronic fatigue and tiredness.

Bladderwrack poultices are good for swollen joints and muscles.

CHINESE MEDICINE: Bladderwrack has a strengthening effect on the Heart. It contains glycosides which influence seratonin production, making it useful in the treatment of depression and sadness.

DOSAGES:

TINCTURE: 1:1, 25% alcohol, 4-8 ml three times daily.
INFUSION: 1:5, 100 ml three times daily or 1-2 tbsp in 1 cup of water, drunk three times daily.
Bladderwrack can also be soaked and eaten.
Poultices for external use.

- The plant is not recommended for children.
- Iodine was in the past produced from bladderwrack.

Menyanthes trifoliata – Gentian family (Gentianaceae)

BOGBEAN

RANGE AND HABITAT: Common throughout Iceland. Grows in fens, ditches and shallow ponds.

PARTS USED MEDICINALLY: Leaves.

GATHERING: Early summer.

ACTIVE SUBSTANCES: Bitter glycosides, including loganin and foliamentin, also flavonoids, saponins, volatile oils, inulin, choline, vitamin C and iodine.

MEDICINAL ACTION: Stimulates digestion, laxative, anti-inflammatory, diuretic and anti-febrile.

USES: Bogbean is useful in treating all sorts of rheumatism, such as muscular rheumatism, rheumatoid arthritis and nerve pain.

Bogbean can also be used for other chronic inflammations, but not for stomach ulcers. The herb stimulates digestion and was formerly thought useful to treat scurvy.

Hot poultices of bogbean leaves can be used externally on swollen joints and muscles.

CHINESE MEDICINE: Bogbean is very drying and has a strengthening effect on Kidneys and Lungs. It is used to treat rheumatism which is accompanied by Dampness and for mucous in the Lungs. It should be avoided where there is Dryness in the body.

DOSAGES:

TINCTURE: 1:5, 25% alcohol, 1-5 ml three times daily.
INFUSION: 1:10, 100 ml three times daily or 1 tsp in 1 cup of water, drunk three times daily. Poultices for external use.

- Smaller doses are required for children.
- Another name for the plant is saddle-pad grass, because the thick turf formed by its rhizomes could be used under a pack saddle.

Carum carvi – Parsley family (Apiaceae)

CARAWAY

MERIDIAN FENNEL, PERSIAN CUMIN

RANGE AND HABITAT: Grows widely in meadows in South Iceland. Elsewhere it is generally only found near farmhouses.

PARTS USED MEDICINALLY: Seeds. Caraway seeds are found in all grocery stores.

GATHERING: Late summer.

ACTIVE SUBSTANCES: Volatile oils, containing for instance carvone and limonene, also tannins, lipids and protein.

MEDICINAL ACTION: Expectorant, antispasmodic, analgesic and antiflatulent, mildly stimulating for the menstrual cycle, stimulates lactation.

USES: Caraway is very useful for flatulence and pain in the digestive tract. It stimulates appetite and for this reason is useful in food for children and adults suffering from chronic low appetite. Caraway is considered helpful to treat all types of colds and coughs and studies have shown that it combats *Helicobacter pylori*.

It stimulates lactation and since the volatile oils are passed into the milk infants benefit from them.

CHINESE MEDICINE: Caraway corrects Liver Qi and strengthens the Stomach.

DOSAGES:

TINCTURE: 1:5, 45% alcohol, 1-3 ml three times daily.
INFUSION: 1:10, 25–50 ml three times daily or ½-1 tsp in 1 cup of water, drunk three times daily. Seeds eaten fresh or dried.

- Smaller doses are required for children.
- Caraway seeds are used as spice, for instance, in bread and soups, and the leaves have also been eaten. It is thought that a 17th-century Icelandic reformer, known as Gísli the Wise, was the first person to grow caraway, in the Fljótshlíð district of South Iceland, around 1660.

Stellaria media – Caryophyllaceae

CHICKWEED

RANGE AND HABITAT: Common everywhere in Iceland. Grows near homes and near farmhouses and in well manured soil. Most people consider this plant a weed.

PARTS USED MEDICINALLY: Entire plant in flower.

GATHERING: All summer long.

ACTIVE SUBSTANCES: Mucilage, saponins, some anti-inflammatory, and various nutrient salts.

MEDICINAL ACTION: Demulcent, vulnerary, refrigerant and astringent The procumbent stalks are thought to stimulate and strengthen the liver and spleen.

USES: Chickweed is used mostly as a skin ointment against all types of inflammation, ulcers and eczema. The herb is especially useful to relieve itching. Chickweed is used internally to treat all types of rheumatism. Infusion of chickweed was thought to stimulate appetite and relieve constipation. The procumbent stalks can be used to treat illnesses caused by liver disorders and for gall stones.

CHINESE MEDICINE: The Chinese use chickweed root, which is sweet and cold and has a strengthening effect on the Stomach and Liver. The root eliminates Heat formed because of deficient Yin energy.

DOSAGES:

TINCTURE: 1:5, 45% alcohol, 4-7 ml three times daily.
INFUSION: 1:10, 100–200 ml three times daily or 1-2 tsp in 1 cup of water, drunk three times daily. The root must be boiled in the same proportions as the infusion of the plant is prepared. Liniment and ointment for external use should be preferably prepared from the fresh plant.

- Smaller doses are required for children.
- Chickweed makes a healthy addition to salads and was traditionally recommended for persons with a weak stomach.

Euphrasia frigida – Figwort family (Scrophulariaceae)

COLD EYEBRIGHT

RANGE AND HABITAT: Common throughout Iceland. Grows in many types of dry soils.

PARTS USED MEDICINALLY: The entire plant in flower, with the exception of the root.

GATHERING: July and August.

ACTIVE SUBSTANCES: Glycosides, including aucubin, tannins, resin and volatile oils.

MEDICINAL ACTION: Astringent, anti-inflammatory, drying mucous (medicine for runny nose) and strengthening mucous membranes.

USES: Eyebright is one of the best herbs available here for persistent runny nose, sinus inflammation and bronchitis. The plant strengthens mucous membranes and dries excessive secretions, and thereby also counteracts gastroenteritis.

Eyebright got its name from its usefulness in curing various infirmities of the eyes. To this end an eyewash can be used of a tincture (thinned with water) or weak infusion made from the plant. Eyebright is recommended, for instance, for conjunctivitis (bloodshot eyes), cataracts and sensitive or sore eyes.

CHINESE MEDICINE: Eyebright is cooling and strengthens the Spleen and Lungs. It helps the Spleen to process energy from food and eliminates Dampness.

DOSAGES:

TINCTURE: 1:5, 45% alcohol, 1-5 ml three times daily.
INFUSION: 1:10, 100 ml three times daily or 1 tsp in 1 cup of water, drunk three times daily. Infusion and tincture as eyewash, see above.

- Smaller doses are required for children.
- Eyebright can be used in ale instead of hops. The ale is said to be tasty and strengthening.

Tussilago farfara – Daisy family (Asteraceae)

COLTSFOOT

RANGE AND HABITAT: Common in urban areas, especially in Reykjavík and the surrounding area. Grows near homes and near farmhouses and developed areas.

PARTS USED MEDICINALLY: Leaves, flower buds and newly opened flowers.

GATHERING: Early summer.

ACTIVE SUBSTANCES: Mucilage (especially in flowers), tannins (especially in leaves), bitter glycosides, zinc, inulin and hormone-like substances.

MEDICINAL ACTION: Sedative and demulcent, expectorant. Used externally, coltsfoot is vulnerary and demulcent.

USES: Coltsfoot is used extensively for all sorts of coughs, as the family name Tussilago (cough) suggests. Coltsfoot is especially useful for dry, hot coughs, spasmodic coughing spells and asthma.

Coltsfoot has always been considered especially effective for children.

Externally the herb can be used in poultices for inflammations and ulcers.

CHINESE MEDICINE: Coltsfoot flowers are bitter and warming and have a strengthening effect on the Lungs. They lead Qi downward and stop coughs. Due to their warming quality, they are especially effective for coughs arising from Cold and Dampness.

DOSAGES:

TINCTURE: 1:5, 45% alcohol, 2-4 ml three times daily.
INFUSION: 1:10, 50-75 ml three times daily or 1 tsp in 1 cup of water, drunk three times daily
Coltsfoot flowers are best fried or heated in honey for dry coughs, 1.5-5 g, twice to three times daily.
Poultices for external use.
The leaves are crushed and softened in boiling water for external use.

- Smaller doses are required for children.
- Recently certain substances have been found in tussilago which are considered to cause liver cancer in rats. Few herbalists have been concerned by this outcome since the plant is not used in anything like the dosage involved in this study.

Pinquicula vulgaris – Bladderwort family (Lentibulariaceae)

COMMON BUTTERWORT

RANGE AND HABITAT: Common throughout Iceland. Grows in peat bogs and marshlands.

PARTS USED MEDICINALLY: Leaves.

GATHERING: Early summer.

ACTIVE SUBSTANCES: Tannins, benzoic and valeric acids, enzymes which curdle milk, and mucilage.

MEDICINAL ACTION: Expectorant and respiratory tonic, antispasmodic and laxative, vulnerary and demulcent.

USES: Common butterwort was formerly used to treat whooping cough successfully. The herb can also be used for dry, tickling coughs and persistent coughing attacks.

The leaves are good for placing on bad sores, inflammation or cracked skin, and have been used to kill lice on the scalp.

DOSAGES:

TINCTURE: 1:5, 25% alcohol, 1-2 ml three times daily.

INFUSION: 1:10, 20-30 ml three times daily or 1/2 tsp in 1 cup of water, drunk three times daily. Poultices and liniment for external use.

- Butterwort was formerly used for making Icelandic skyr and is therefore also called "curdle herb". The plant is carnivorous, catching and digesting small insects.

Senecio vulgaris – Daisy family (Asteraceae)

COMMON GROUNDSEL

RANGE AND HABITAT: A common weed in cultivated land.

PARTS USED MEDICINALLY: The whole plant, with the exception of the root.

GATHERING: All summer long.

ACTIVE SUBSTANCES: The alkaloids senecine and seniocine, inulin, resins and potassium.

MEDICINAL ACTION: Sudorific, laxative, diuretic, cooling and analgesic.

USES: A weak infusion of common groundsel has a mildly laxative effect and was formerly used to eliminate parasites from children's digestive systems. The herb has a cooling effect on the digestive system and is considered useful for treating gallbladder complaints and spasms.

Poultices of common groundsel are useful to treat swollen and tender parts of the body, such as sore joints. The herb helps heal dermal sores.

DOSAGES:

TINCTURE: 1:5, 25% alcohol, 1-2 ml three times daily.
INFUSION: 1:10, 20-25 ml three times daily or 1/2 tsp in 1 cup of water, drunk three times daily. Poultices and infusion for external use.

- The plant is not recommended for children.

WARNING!
The alkaloids in common groundsel can cause a miscarriage and pregnant women should absolutely avoid using the herb.

Calluna vulgaris – Heather family (Ericaceae)

COMMON HEATHER

LING, HEATHER

RANGE AND HABITAT: Grows in moorlands and on slopes, common throughout most of Iceland, although rare in the area between Snæfellsnes peninsula and Skagafjörður.

PARTS USED MEDICINALLY: Flowering tips.

GATHERING: August and September.

ACTIVE SUBSTANCES: Citric acid, fumaric acid, the alkaloid ericodin, tannins, arbutin, flavonoids and glycosides.

MEDICINAL ACTION: Diuretic and urinary disinfectant, anti-inflammatory, particularly antirheumatic, vulnerary and mildly sedative.

USES: Heather is used mainly as a urinary antiseptic, especially to treat cystitis. Heather is helpful for various rheumatic conditions, especially gout. Persons suffering from insomnia can benefit from drinking a heather infusion in the evening.

A liniment made from heather tips is very useful as a massage against arthritic pain. The liniment relieves both the pain and inflammation.

CHINESE MEDICINE: Heather is cooling and strengthens Liver Yin. It is nourishing for bones and joints, eliminates Heat and is nourishing for women in menopause who suffer from insomnia, nightly heat flushes and dryness of mucous membranes.

DOSAGES:

TINCTURE: 1:5, 25% alcohol, 1-5 ml three times daily.
INFUSION: 1:10, 100 ml three times daily or 1 tsp in 1 cup of water, drunk three times daily. Liniment and hot poultices for external use.

- Smaller doses are required for children.
- As heather was traditionally believed to deter mice and rats, it was commonly gathered and stored indoors.

Galeopsis tetrahit – Mint family (Lamiaceae)

COMMON HEMP NETTLE

RANGE AND HABITAT: In uncultivated fields and in gardens.

PARTS USED MEDICINALLY: Leaves.

GATHERING: Summer prior to flowering.

ACTIVE SUBSTANCES: Mucilage, tannins, oils and saponins.

MEDICINAL ACTION: Astringent, diuretic and expectorant.

USES: Hemp nettle is especially effective as an expectorant and is used extensively for bad coughs. The herb is thought to be effective against anaemia and other disorders of the blood. Hemp nettle has long been used to strengthen the spleen.

CHINESE MEDICINE: Hemp nettle strengthens Spleen and helps it produce energy from food. Because of the effect of hemp nettle on the Spleen it is used for diabetes and internal haemorrhaging.

DOSAGES:

TINCTURE: 1:5, 25% alcohol, 1-5 ml three times daily.
INFUSION 1:10, 100 ml three times daily or 1-2 tsp in 1 cup of water, drunk three times daily.
Hemp nettle soup, made from young leaves boiled for 15-20 minutes, is very strengthening and purging for the blood.

- Smaller doses are required for children.

Juniperus communis – Cypress family (Cupressaceae)

COMMON JUNIPER

RANGE AND HABITAT: Common throughout most of Iceland. Grows in dry soils, lava, brushland and moorland.

PARTS USED MEDICINALLY: Berries and new foliage (needles).

GATHERING: The berries are gathered in the autumn of the second or third year when they have turned blue-black, while the needles are gathered in the spring.

ACTIVE SUBSTANCES: Volatile oils, containing for instance pinene and camphene, bitter principles, including juniperine, polophyllotoxin (anti-tumour agent) flavonoids, tannins and sugars.

MEDICINAL ACTION: Diuretic, antibacterial, especially in the urinary organs, anti-rheumatic, improves digestion, anti-flatulent and stimulates contraction of the womb.

USES: The berries and leaves have a similar effect, although the berries are stronger. They are used mostly for infections of the urinary tract, but are also useful in treating all types of rheumatism, especially gout.

Juniper berries have been used extensively to strengthen the stomach and improve digestion, not least where stomach acidity is insufficient. They can also counteract spasms and flatulence.

Oil prepared from juniper berries can be used on rheumatic muscles and joints, and is also considered useful for inflamed nerves and muscles.

CHINESE MEDICINE: Juniper berries and needles are warming and strengthen Kidney Qi and Liver.

DOSAGES:

TINCTURE: 1:5, 45% alcohol, 1-2 ml three times daily.
INFUSION of berries and/or leaves: 1:10, 25–50 ml three times daily or 3 berries: 1 cup of water, drunk three times daily.
Oil of juniper for external use.

WARNING!

Some substances in juniper berries (and needles) can irritate the kidneys, and therefore persons with kidney ailments should not use juniper. Juniper should not be used for longer than 6 weeks at a time. Pregnant women should not use this herb at all. The plant is not recommended for children.

Polygonum aviculare – Knotweed family (Polygonaceae)

COMMON KNOTGRASS

RANGE AND HABITAT: The whole plant, with the exception of the root.

GATHERING: All summer long.

ACTIVE SUBSTANCES: Silicic acid, anthraquinones, tannins, saponins and mucilage.

MEDICINAL ACTION: Astringent, diuretic, staunches bleeding, expellant of stones and sand in the urinary tract.

USES: Knotgrass is used mainly for internal haemorrhaging, especially in the alimentary tract, and is especially effective against diarrhoea. Knotgrass can be used for stones in the urinary tract, in which case the herb must be taken regularly for a long period. Crushed knotgrass inhaled through the nose is effective in stopping nosebleeds.

Poultices and ointments of knotgrass can be used on suppurating sores.

CHINESE MEDICINE: Knotgrass is cooling and has the greatest effect on the Bladder. It is used against infections and inflammation of the bladder and urinary tract caused by Moist Heat. Knotgrass expels parasites from the body and stops itching. It is therefore used for various parasitical and worm infections which cause dermal itching and irritation, as well as for worms in the intestinal tract.

DOSAGES:

TINCTURE: 1:5, 25% alcohol, 1-3 ml three times daily.
INFUSION: 1:10, 25-50 ml three times daily or 1/2-1 tsp in 1 cup of water, drunk three times daily. Ointment, poultices and crushed herb for external use.

- Smaller doses are required for children.
- Formerly, the plant was believed to retard the growth of children.
- The species name *aviculare* is from the Latin aviculus, a diminutive of avis (a bird), as hens and small birds are very fond of knotgrass seeds.
- The seeds were used for human consumption and likened to buckwheat.

WARNING!
Always seek a physician's advice in the case of internal haemorrhage before attempting treatment with herbs.

Hippuris vulgaris – Plantain family (Plantaginaceae)

COMMON MARE'S TAIL

RANGE AND HABITAT: Grows in lakes and wetlands. Common everywhere in Iceland.

PARTS USED MEDICINALLY: The whole plant, with the exception of the root.

GATHERING: All summer long.

ACTIVE SUBSTANCES: The plant has not been researched extensively, but is known to contain saponins, silica compounds and bitter principles.

MEDICINAL ACTION: Stops internal and external haemorrhaging, strengthens the lungs and liver and is thought to help with kidney stones.

USES: Mare's tail is very useful as to staunch bleeding and has mostly been used for this purpose. It is considered especially good for haemorrhaging in the lungs, digestive tract and kidneys.

Externally the plant is placed on sores which are slow to heal; powder from the plant can be inhaled to stop nose bleeds.

CHINESE MEDICINE: Mare's tail strengthens the Spleen and especially its function of holding Blood in the veins.

DOSAGES:

TINCTURE: 1:5, 25% alcohol, 2-4 ml three times daily.

INFUSION: 1:10, 25-50 ml three times daily or 1/2-1 tsp in 1 cup of water, drunk three times daily. Poultices and powder of the herb for external use.

- Smaller doses are required for children.
- No attempt should be made to staunch internal haemorrhaging without seeking the advice of a physician for the ailment.

Plantago major – Plantain family (Plantaginaceae)

COMMON PLANTAIN

RANGE AND HABITAT: Found in many areas of the country, but least common in the northeast and east of Iceland. Grows in geothermal areas and as a stray by buildings and along the roadside.

PARTS USED MEDICINALLY: Leaves.

GATHERING: Early summer.

ACTIVE SUBSTANCES: Glycosides, including aucubin, also mucilage, acids, tannins, vitamin C, silica compounds and the enzymes emulsin and invertin.

MEDICINAL ACTION: Astringent, demulcent and diuretic, also expectorant and staunches bleeding.

USES: Plantain is used mostly for infections and inflammation of the urinary tract; the herb is especially effective if there is blood in the urine. Plantain is good for scurvy because of its rich vitamin content and can also be used to treat bronchitis and dry, tickling cough. Plantain is also used to treat wounds that are slow to heal, and was called "Soldier's herb" referring to this use. Suppositories containing the herb are used to treat haemorrhoids.

CHINESE MEDICINE: Plantain seeds are sweet and cold and have a strengthening effect on the Bladder, Kidneys, Liver and Lungs. They are diuretic and counteract Heat and are used for all types of oedema and urinary infections. The seeds cleanse the eyes and are used for eye ailments caused by lack of Liver and Kidney Yin (such as dry eyes and cataracts) or Liver Heat (such as red, swollen or sore eyes). The seeds also eliminate Lung-Phlegm and stop coughs, and are used to treat Lung-Heat and Dampness.

DOSAGES:

TINCTURE: 1:5, 45% alcohol, 2-4 ml three times daily.
INFUSION: 1:10, 50-75 ml three times daily or 1 tsp in 1 cup of water, drunk three times daily. The seeds are heated in white wine when used for Kidney Yin deficiency, 4-6 g, three times daily. For other purposes the seeds are either dried at low heat in a pan and eaten or a tincture or infusion is prepared from them. Poultices, liniment, crushed leaves and suppositories for external use.

- **Smaller doses are required for children.**

Polypodium vulgare – Fern family (Polypodiaceae)

COMMON POLYPODY

RANGE AND HABITAT: Grows widely in the south of Iceland, rare elsewhere. Grows in shaded locations, often in lava fields and rocky areas.

PARTS USED MEDICINALLY: Rhizome.

GATHERING: Spring and autumn.

ACTIVE SUBSTANCES: Mannitol, volatile oils, sugars, glycyrrhizin, tannins and saponins, including polypodine.

MEDICINAL ACTION: Strengthening, stimulates gall excretion and other liver functions, laxative and expectorant.

USES: Common polypody is mostly used as a mild laxative. The rhizhome has a positive effect on persistent skin conditions and is considered useful against mycosis, both dermal and in the digestive system.

It is also useful to expel persistent phlegm from the respiratory tract and is often used to treat hoarseness and dry coughs.

CHINESE MEDICINE: Common polypody is warming and drying and strengthens the Liver and Gallbladder.

DOSAGES:

TINCTURE: 1:5, 25% alcohol, 1-2 ml three times daily.
DECOCTION: 1:10, 20-30 ml three times daily or 1/2 tsp in 1 cup of water, drunk three times daily. Boil the rhizhome until the water has become thick and syropy.

- Smaller doses are required for children.
- A decoction of polypody was formerly used against the cold, and thus its Icelandic name of "cold herb".

WARNING!

Large doses of common polypody are very strongly laxative.

Cochlearia officinalis – Mustard family (Brassicaceae)

COMMON SCURVYGRASS

RANGE AND HABITAT: Found widely on cliffs and rocks on tidal beaches and bird cliffs, and also in inland areas.

PARTS USED MEDICINALLY: Leaves and roots.

GATHERING: Early summer.

ACTIVE SUBSTANCES: Vitamin C and other vitamins, bitter principles, tannins, glycosides and a strong oil containing, among other things, sulphur.

MEDICINAL ACTION: Roborant, diuretic and sudorific, stimulates digestion and purges the blood. Scurvygrass leaves, crushed and placed fresh on a wound, make a very good vulnerary.

USES: Long before vitamins became known, common people in Iceland cured scurvy and various other ailments with scurvygrass. Today scurvygrass is used mainly for rheumatism, oedema and various skin conditions, due to its purging effect on the blood.

It is useful to prepare a strengthening and cleansing tincture of the herb for use in winter.

DOSAGES:

TINCTURE: 1:5, 25% alcohol, 1-5 ml three times daily.
INFUSION of leaves and decoction of roots: 1:10, 100 ml three times daily or 1 tsp in 1 cup of water, drunk three times daily. Fresh leaves are best for herbal medicine, as vitamin C is mostly lost in drying.

- Smaller doses are required for children.
- The roots are eaten fresh or stewed and the leaves, picked after flowering, are good in salads and soups and in sandwiches.

Potentilla anserina – Rose family (Rosaceae)

COMMON SILVERWEED

GOOSEWORT, SILVERWEED CINQUEFOIL

RANGE AND HABITAT: Common throughout lowland Iceland. Grows in sandy soil.

PARTS USED MEDICINALLY: The whole plant, with the exception of the root.

GATHERING: Early summer.

ACTIVE SUBSTANCES: Tannins, tormentol, bitter principles and flavonoids.

MEDICINAL ACTION: Astringent, antispasmodic and anti-inflammatory.

USES: Silverweed is very useful to treat inflammation of the digestive tract and is used extensively for cramps and diarrhoea. It is also considered useful for uterine cramps and pain.

Like other astringent herbs, silverweed is useful as a gargle for an inflamed and sore mouth and throat, in both children and adults. It strengthens the gums and periodontium. Silverweed is also used as a rinse for sores and discharges from the vagina.

CHINESE MEDICINE: Silverweed is strengthening for the Liver and corrects Liver Qi.

DOSAGES:

TINCTURE: 1:5, 25% alcohol, 2-5 ml three times daily.
INFUSION: 1:10, 50-100 ml three times daily or 1/2-1 tsp in 1 cup of water, drunk three times daily.
Infusion and tincture as mouthwash.

- Smaller doses are required for children.
- Formerly a poultice of crushed silverweed mixed with honey was placed on the stomach to relieve colic. The roots were formerly eaten, either ground or stewed in water and milk, as referred to in the folk verse, “children had and deed, roots and silverweed”.

Rumex acetosa – Polygonaceae

COMMON SORREL

RANGE AND HABITAT: Grows in many types of dry soils, especially grasslands and hayfields. Very common throughout Iceland.

PARTS USED MEDICINALLY: Leaves.

GATHERING: Early summer.

ACTIVE SUBSTANCES: Oxalic acid, anthraquinone, sugars, tannins and vitamins, including vitamins A, B and C.

MEDICINAL ACTION: Diuretic, laxative, anti-inflammatory and cooling.

USES: Common sorrel is considered useful to treat oedema, especially if caused by liver malfunction which the herb is thought to stimulate and strengthen. It is also good for poor appetite, scurvy, constipation and haemorrhoids.

Fresh juice of the leaves, diluted with vinegar, is an excellent lotion for skin disorders, such as furuncles, mycosis, itchy eczema and neoplasms. For best effect the herb can be taken orally at the same time.

WARNING!

Persons suffering from rheumatoid arthritis, especially gout, should not use common sorrel because of its high acidity. The same applies to persons with kidney stones and excess stomach acid. Overdoses of the herb can result in kidney failure.

DOSAGES:

TINCTURE: 45% alcohol, 1-4 ml three times daily.

INFUSION: 1:10, 40-60 ml three times daily or ½ tsp in 1 cup of water, drunk three times daily.

Pressed juice from the leaves for external use.

- Smaller doses are required for children.
- Wood sorrel (*Oxyria dygina*) has characteristics and effects similar to those of common sorrel. Common sorrel is often confused with Sheep's sorrel (*Rumex acetosella*), a closely related, smaller herb.

Veronica officinalis – Plantain family (Plantaginaceae)

COMMON SPEEDWELL

RANGE AND HABITAT: Common through much of Iceland. Grows in brush and coppices.

PARTS USED MEDICINALLY: The dried plant in flower, with the exception of the root.

GATHERING: Late summer.

ACTIVE SUBSTANCES: Tannins, the glycoside aucubin, volatile oils, saponins and bitter principles.

MEDICINAL ACTION: Sudorific and diuretic, expectorant and heals externally.

USES: Common speedwell is used less for medicinal purposes today than formerly, as the herb has a very mild effect. It was traditionally used for many ailments, such as scurvy, lack of appetite, anorexia, kidney stones, rash and irritated skin and even to relieve cataracts.

This plant was also thought to help dry cough and purge the blood and poultices were used externally to heal sores.

DOSAGES:

INFUSION: 1:10, 50 ml once or twice daily or 1 tsp in 1 cup of water, drunk once to twice daily. Poultices for external use.

- Smaller doses are required for children.
- Among other names for the plant are groundheal.

Sedum acre – Pink family (Crassulaceae)

COMMON STONECROP

RANGE AND HABITAT: Common everywhere in Iceland. Grows in gravelly, rocky and sandy areas.

PARTS USED MEDICINALLY: The fresh plant before flowering, with the exception of the root.

GATHERING: Early summer.

ACTIVE SUBSTANCES: Alkaloids, including sedamine, as well as glycosides and mucilage.

MEDICINAL ACTION: Irritates the skin, causing a rash and even blisters.

USES: Stonecrop is only used externally. The herb can cause undesirable side effects, e.g. vomiting or diarrhoea, if taken internally.

Stonecrop is considered very useful to treat warts and corns, on which a poultice is placed three to four times daily. The herb is also useful for sores which are slow to heal and to stimulate dermal blood flow.

A weak infusion or liniment of the plant is best for sores.

DOSAGES:

The fresh herb, crushed for external use.
INFUSION for external use.
Liniment for external use.

- Stonecrop has little or no effect after drying.
- Traditionally, decoction of stonecrop was thought effective for earache.

Drosera rotundifolia – Sundew family (Droseraceae)

COMMON SUNDEW

RANGE AND HABITAT: Grows in marshlands and fens. Very rare, but found mainly in the west and north of Iceland.

PARTS USED MEDICINALLY: The whole plant, with the exception of the root.

GATHERING: Early summer.

ACTIVE SUBSTANCES: Naphthoquinone, droserone, flavonoids, tannins and other organic acids.

MEDICINAL ACTION: Demulcent, antispasmodic and expectorant, considered antifebrile.

USES: Sundew relieves convulsive coughing spells and has been used successfully to treat whooping cough. It is also useful for asthma and dry, tickling coughs.

Sundew contains antibacterial substances which are thought effective against various common bacteria causing colds and bronchitis. For this reason, it is recommended that the herb be used when a cold begins, especially if accompanied by a dry, irritating cough, to prevent subsequent bacterial infection. Sundew can also be used for gastritis and ulcers of the digestive tract.

Externally the fresh juice is used on warts and corns.

CHINESE MEDICINE: Sundew is cooling and strengthens the Stomach and Lungs.

DOSAGES:

TINCTURE: 1:5, 45% alcohol, ½–1 ml three times daily.
INFUSION: 1:10, 25-50 ml three times daily or 1/2 tsp in 1 cup of water, drunk three times daily.
Fresh juice for external use.

- Smaller doses are required for children.
- Because of its rarity in Iceland, it is recommended that people cultivate this plant for their own use.
- An insectivorous plant, sundew has tiny glandular hairs on its leaves with which it captures small insects.
- Traditionally sundew was believed to eliminate freckles.

WARNING!

The herb contains strongly irritating substances so that large doses can cause undesirable side effects.

Oxalis acetosella – Oxalidaceae

COMMON WOOD SORREL

RANGE AND HABITAT: Extremely rare and completely protected in Iceland. Grows in cold, shady locations.

PARTS USED MEDICINALLY: The leaves, which are available in health food stores.

GATHERING: Early summer.

ACTIVE SUBSTANCES: Potassium bioxalate, vitamin C, mucilage and oxalic acid.

MEDICINAL ACTION Diuretic, cooling, astringent, purges the blood and strengthening for the stomach.

USES: An infusion of wood sorrel is good as a drink and to reduce a high fever. The plant is useful to treat excessive mucous secretion in both the urinary and digestive system. Wood sorrel was formerly used to improve digestion; it stimulates appetite mildly and reduces nausea.

CHINESE MEDICINE: Wood sorrel is very strengthening for Spleen Qi and therefore helps to produce energy from food.

DOSAGES:

TINCTURE: 1:5, 25% alcohol, 1-5 ml three times daily.
INFUSION: 1:10, 100 ml three times daily or 1 tsp in 1 cup of water, drunk three times daily.

WARNING!

As wood sorrel is high in oxalic acid, persons suffering from rheumatism or kidney and gallbladder stones should not use the plant.

- Smaller doses are required for children.
- Because of its rarity in Iceland, it is recommended that people cultivate this plant for their own use.

Elymus repens – Grass family (Poaceae)

COUCH GRASS

RANGE AND HABITAT: Around farms and in neglected gardens throughout Iceland, it is regarded as among the worst of weeds.

PARTS USED MEDICINALLY: Rhizomes and stolons.

GATHERING: Spring and autumn.

ACTIVE SUBSTANCES: Mannitol, silica compounds, mucilage, latex, vanillin, saponins, agropyren, which is an antibacterial substance, iron and other minerals.

MEDICINAL ACTION: Vulnerary, demulcent and anti-bacterial for urinary organs. Diuretic and relieves pain and cramps of the urinary tract.

USES: Couch grass is used extensively to treat urinary ailments, e.g. cystitis, prostatitis, nephritis and discomfort from kidney stones and gravel. The plant has a prompt effect and is completely harmless, even for children.

CHINESE MEDICINE: Couch grass is cooling and strengthens Kidney Yin and eliminates Heat from the Bladder.

DOSAGES:

TINCTURE: 1:5, 25% alcohol, 1-5 ml three times daily.
INFUSION: 1:10, 100 ml three times daily or 1 tbsp in 1 cup of water, drunk three times daily.

- Smaller doses are required for children.

Thymus praecox – Mint family (Lamiaceae)

CREEPING THYME

PRUNELLA, ALL-HEAL, HEART OF THE EARTH

RANGE AND HABITAT: Common throughout Iceland. Grows on gravel flats and dry moors and slopes.

PARTS USED MEDICINALLY: The entire plant in flower, with the exception of the root.

GATHERING: All summer long.

ACTIVE SUBSTANCES: Volatile oils, including, for instance, thymol and carvacrol, as well as tannins, saponins, resins, flavonoids and bitter principles.

MEDICINAL ACTION Antiseptic, expectorant, antispasmodic and eliminates flatulence.

USES: Thyme is most commonly used for flu and colds, especially bronchitis and other lung ailments, where it works as an antiseptic and expectorant. Thyme is also very effective against various digestive complaints, such as gastric and intestinal inflammation, and is thought to eliminate helicobacter infections which cause gastritis and stomach ulcers. Thyme is used, often together with other herbs, to relieve cramps of the digestive tract.

A strong infusion of the plant, drunk 4-6 times a day, for 2-3 days at a time, is recommended to help people stop drinking alcoholic beverages. However, this would likely have to be repeated several times with intervals in between. Many people believe that an infusion of Thyme will alleviate a hangover.

Thyme also makes a good mouthwash to relieve sore throat or mouth sores. Poultices are recommended for swollen joints and muscles.

CHINESE MEDICINE: Thyme is cooling and an expectorant and has a strengthening effect on the Liver and Stomach.

DOSAGES:

TINCTURE: 1:5, 45% alcohol, 2-4 ml three times daily.

INFUSION: 1:10, 40-60 ml three times daily or 1 tsp in 1 cup of water, drunk three times daily.

Infusion and diluted tincture as mouthwash.

Poultices for external use.

- Smaller doses are required for children.
- Thyme is a very popular spice (timian).
- According to an old Icelandic herbal, thyme purges the blood, is sedative and dilutes secretions.

Taraxacum officinale – Daisy family (Asteraceae)

DANDELION

RANGE AND HABITAT: Common everywhere in Iceland. Grows in grassland, near homes and farmhouses and in the mountains.

PARTS USED MEDICINALLY: Roots and leaves.

GATHERING: Harvested before flowering.

ACTIVE SUBSTANCES: Bitter glycosides, including taraxacin, phytohormones, including cytosterol, taraxasterol and taraxerin, also inulin and other sugars, tannins, wax, various vitamins and minerals, e.g. potassium in quantity.

MEDICINAL ACTION: Diuretic, strengthens the liver and digestive organs, laxative and stimulates gall secretion.

USES: Dandelion is a very important medicinal herb. Many people mix the roots and leaves together to gain most advantage of the effects of both.

The leaves, which are very nutritious, have little effect on the liver, but are diuretic and high in potassium. They are therefore used extensively for oedema, especially if caused by a weak heart.

DOSAGES:

TINCTURE: 1:5, 25% alcohol, 5-10 ml three times daily.
INFUSION of leaves and decoction of roots: 1:10, 100–200 ml three times daily or 1-2 tsp in 1 cup of water, drunk three times daily. Dandelion milk for external use.

The roots are used for all liver and gall bladder complaints, e.g. jaundice, and also for slow bowels, insomnia and depression. The roots are very useful to regain strength after a lengthy course of allopathic medicines or protracted alcohol consumption. The roots are also effective against oedema caused by liver malfunction.

Dandelion milk can be used on warts and corns.

CHINESE MEDICINE: The Chinese use the entire plant including the root, gathered when the flowers are about to open. Dandelion is bitter, sweet and cooling, affecting the Liver and Stomach. It eliminates Heat, especially Liver-Heat, manifesting in swollen and red eyes, and works against furuncules and lumps, especially hard ones in the breasts, small intestine or colon. Dandelion stimulates lactation and is used to eliminate Damp Heat and for urinary symptoms caused by Heat or Dampness.

- Smaller doses are required for children.
- Dandelion is a completely harmless herb, even in large doses.
- A tasty wine can be brewed from the flowers and the leaves are considered tasty in salad.
- Formerly, decoction of dandelion leaves was used to bathe the face to improve the appearance and the roasted roots were used as a coffee substitute.

Palmaria palmata – Palmariaceae

DULSE

RANGE AND HABITAT: Grows on tidal beaches between the limits of high and low tide. Most common along the west and south coast.

PARTS USED MEDICINALLY: Entire plant.

GATHERING: Late summer.

ACTIVE SUBSTANCES: Protein, vitamins A and B, glycosides and various minerals, e.g. dulse is rich in iodine.

MEDICINAL ACTION/USAGE: Dulse is very nutritious, laxative, diuretic and sudorific. It stimulates both appetite and thirst. Dulse is useful for motion sickness, i.e. nausea and dizziness, and as a hangover treatment, since it is very salty.

CHINESE MEDICINE: Like bladderwrack, dulse affects the Heart. It contains glycosides which influence seratonin production, making it useful to treat depression and low mood.

DOSAGES:
Dulse is dried and can be eaten freely on its own or mixed with other foods.

- An 18th-century Icelandic herbalist wrote of dulse, "persons who have upset the functioning of their stomach and blood regulation through excessive drink can benefit from taking dulse before eating or drinking the following day".

Betula pubescens – Birch family (Betulaceae)

DWARF BIRCH

RANGE AND HABITAT: Common throughout most of Iceland, forming small woods and brushlands at altitudes up to 400m.

PARTS USED MEDICINALLY: New leaves, bark of newest branches and sap.

GATHERING: Spring. The bark must be cut into the wood before leaves appear to tap the sap (birchwater). It is then allowed to drip into a container. Effort should be made to make the smallest cut as possible in each tree.

ACTIVE SUBSTANCES: Saponins, volatile oils containing, for instance, betulin, resins, flavonoids, tannins and bitter principles.

MEDICINAL ACTION: Diuretic, anti-inflammatory, sudorific, stimulates the liver and purges the blood.

USES: Birch is used mostly for all types of rheumatic complaints, especially for kidney disorders. It greatly strengthens the kidneys, whether they are plagued by infection or other complaints. Birch is especially useful with other herbs which affect rheumatism directly, such as meadow sweet and bogbean. Birch is sometimes effective on psoriasis-eczema and rheumatism. Birch reduces hypertension and oedema. Ointments from decoctions of birch leaves and bark are used for external healing.

CHINESE MEDICINE: Birch strengthens Kidney Qi and Yin. It has a positive effect on bones and joints and eliminates Heat. It also strengthens the Liver and helps renew and purge the Blood.

DOSAGES:

TINCTURE: 1:5, 25% alcohol, 1-5 ml three times daily.
INFUSION of the leaves or decoction of bark: 1:10, 100 ml three times daily or 1 tsp in 1 cup of water, drunk three times daily.
BIRCHWATER in 20% alcohol (to lengthen storage life) 1:3, 5-10 ml three times daily.

- Smaller doses are required for children.
- Birchwater can also be used to brew good wine.

Gentianella campestris – Gentian family (Gentianaceae)

FIELD GENTIAN

RANGE AND HABITAT: Grows in dry meadows and on slopes. Fairly common almost everywhere in Iceland.

PARTS USED MEDICINALLY: Root.

GATHERING: Spring and autumn.

ACTIVE SUBSTANCES: Bitter glycosides and flavonoids.

MEDICINAL ACTION: Roborant, stimulates digestion by increasing production of digestive fluids, such as stomach acid, and stimulates menstruation.

USES: Like other bitter herbs, field gentian is useful to treat lack of appetite and other symptoms resulting from digestive disturbances and liver weakness.

Field gentian strengthens persons who have been struggling with long illnesses and are failing to recover.

Field gentian and other bitter herbs are used as part of treatment for gallstones.

CHINESE MEDICINE: Field gentian is a bitter, cold and acidic herb, affecting the Gallbladder, Liver and Stomach. The root eliminates Wind-Dampness and relaxes tendons, so that it is used to treat arm and leg cramps caused by Dampness or Wind. It is used to treat jaundice, especially in infants, and because of its effects on digestion to relieve bowel impactions.

DOSAGES:

TINCTURE: 1:5, 45% alcohol, 1-2 ml three times daily.
DECOCTION: 1:10, 25-50 ml three times daily or 1/2 tsp in 1 cup of water, drunk three times daily.
When the herb is used to stimulate appetite it should be taken ½–1 hour before a meal.

- Smaller doses are required for children.
- Formerly the herb was used for a variety of ailments, including weak heart, flatulence, intestinal parasites and rheumatism.

Equisetum arvense – Horsetail family (Equisetaceae)

FIELD HORSETAIL

RANGE AND HABITAT: Common throughout Iceland. Grows in peat bogs, woods and on cultivated land.

PARTS USED MEDICINALLY: The whole plant, with the exception of the root. The spore-bearing buds which top the stems (devil's feet) are considered the best part.

GATHERING: Spring.

ACTIVE SUBSTANCES: Silica compounds, saponins, alkaloids, bitter principles, vitamin C and plenty of minerals, especially potassium.

MEDICINAL ACTION: Astringent, stops bleeding, diuretic, strengthening and vulnerary for lungs and kidneys, removes various toxins from the body and strengthens the immune system.

USES: Field horsetail is useful to treat most illnesses affecting the kidneys and upper urinary tract. The plant strengthens urinary organs, e.g. following the use of strong drugs such as antibiotics and cancer therapies. It reduces oedema and immoderate nightly urine production. Field horsetail is vulnerary for the lungs, and encourages blood clotting. It has therefore been used extensively against haemorrhaging, both internal and external.

It is useful to place poultices on bad wounds, e.g. those which bleed heavily.

This herb was investigated extensively in the former Soviet Union, where it is maintained that it purges lead from the body.

FOR WOMEN: Field horsetail strengthens all organs in the pelvic girdle and has been used successfully to treat endometriosis and chronic uteritis and ovaritis.

CHINESE MEDICINE: Field horsetail, like most other horsetails, is sweet, acrid and neutral and affects Kidneys and Liver. It eliminates Wind-Heat and stops bleeding. The herb is especially good with Wind-Heat that troubles the eyes, causing redness, pain, inflammation, cloudy vision and excess tear secretion.

DOSAGES:

TINCTURE: 1:5, 25% alcohol, 1-5 ml three times daily.
INFUSION: 1:10, 100 ml three times daily or 1-2 tsp in 1 cup of water, drunk three times daily. Poultices and infusion for external use.

- Smaller doses are required for children.
- Meadow horsetail, also known as Shade horsetail (*Equisetum pratense*) is thought to have similar effects as common horsetail. Many people have difficulty in distinguishing between these two types.

Epilobium angustifolium – Onagraceae

FIREWEED

WILLOWHERB

RANGE AND HABITAT: Grows on cliffs and woodlands, as well as near farms, often in dense patches. Relatively rare.

PARTS USED MEDICINALLY: The whole plant, with the exception of the root.

GATHERING: All summer long.

ACTIVE SUBSTANCES: Mucilage, tannins and various salts.

MEDICINAL ACTION: Astringent, demulcent and vulnerary, antispasmodic.

USES: Fireweed is recommended for inflammation of the digestive tract. Because of its characteristics, the herb is especially useful for ailments accompanied by spasms or diarrhoea.

Used externally, fireweed is vulnerary and demulcent. Fireweed is a very mild herb which can therefore be used by children and adults alike.

DOSAGES:

TINCTURE: 1:5, 25% alcohol, 1-5 ml three times daily.
INFUSION: 1:10, 100 ml three times daily or 1 tsp in 1 cup of water, drunk three times daily.
Liniment and poultices for external use.

- Smaller doses are required for children.
- Young shoots can be used in the same manner as asparagus and the leaves as greens.

Myosotis arvensis – Forget-me-not family (Boraginaceae)

FORGET-ME-NOT

RANGE AND HABITAT: Common almost everywhere in Iceland. Grows in many types of dry soils, especially close to settled areas.

PARTS USED MEDICINALLY: The whole plant, with the exception of the root.

GATHERING: All summer long.

ACTIVE SUBSTANCES: The plant has not been researched extensively, but is known to contain both tannins and mucilage.

MEDICINAL ACTION: Astringent, purges the blood, demulcent and vulnerary.

USES: Forget-me-not is considered beneficial to the lungs and was often used to treat many types of pulmonary disorders. The herb is also used on minor sores and burns.

DOSAGES:

TINCTURE: 1:5, 25% alcohol, 1-2 ml three times daily.
INFUSION: 1:10, 20-30 ml three times daily or 1/2 tsp in 1 cup of water, drunk three times daily. Poultices and infusion for external use.

- Smaller doses are required for children.
- The plant clings to clothing and children often use it as an ornament.

GARDEN ANGELICA

HOLY GHOST, WILD CELERY

RANGE AND HABITAT: Grows in fertile hollows, rocky cliffs, and the banks of lakes, rivers and streams. Fairly common throughout Iceland.

PARTS USED MEDICINALLY: Entire plant.

GATHERING: The roots must be gathered in the autumn of the plant's first year. Leaves are gathered in the summer and seeds when fully ripe.

ACTIVE SUBSTANCES: Volatile oils, containing for instance pellandrin and pinin. Oils of the seeds contain, in addition, methylethylacetic acids and hydroxymyristic acids. The entire plant contains angelic resins, angelic acids, coumarin substances, bitter principles and tannins.

MEDICINAL ACTION: Strengthening and warming for the digestive organs, diuretic, antiflatulent, antispasmodic, expectorant, stimulates digestion and eliminates stress.

USES: Garden angelica is especially good for persons recovering from serious illness, suffering from poor appetite, constipation etc. Garden angelica is used to treat digestive disturbances, such as cramps and flatulence in the digestive organs, and liver ailments. The plant is very useful as an expectorant, and can thus be used for bronchitis and pleurisy. Garden angelica is often recommended for asthma and other children's lung ailments.

The seeds are thought to strengthen the immune system and are therefore useful for cancer.

CHINESE MEDICINE: Common angelica strengthens the Liver and corrects Liver Qi.

DOSAGES:

TINCTURE: 1:2, 45% alcohol, 2-3 ml three times daily.
INFUSION of leaves and decoction of roots and seeds: 1:10, 25-50 ml three times daily or 1/2 tsp in 1 cup of water, drunk three times daily.

- Smaller doses are required for children.
- Wild angelica (*Angelica sylvestris*) has an effect similar, although weaker, to common angelica and can therefore be used in the same manner.
- Common angelica has always been held in high regard in Iceland as a medicinal herb.

WARNING!

Common angelica stimulates menstruation and therefore pregnant women should not use the herb, especially during the first months of pregnancy.

Parnassia palustris – Bittersweet family (Celastraceae)

GRASS OF PARNASSUS

NORTHERN GRASS OF PARNASSUS, BOG STAR

RANGE AND HABITAT: Grows in peat bogs and wetlands, very common everywhere in Iceland.

PARTS USED MEDICINALLY: The whole plant, with the exception of the root.

GATHERING: Early summer.

ACTIVE SUBSTANCES: Tannins, resins, mucilage and various minerals.

MEDICINAL ACTION: Astringent, roborant, vulnerary and demulcent.

USES: Grass of Parnassus is considered to strengthen the liver especially, and is therefore used extensively for persistent hepatitis and digestive tract ailments, haemorrhoids and persistent diarrhoea. The plant has also been used to treat depression, like so many others which strengthen the liver. Grass of Parnassus is said to have a sedative and strengthening effect on the heart and for this reason has been used for heart arrhythmias.

The herb is vulnerary and therefore used on wounds which are slow to grow, such as burns.

CHINESE MEDICINE: Grass of Parnassus is warming and astringent, strengthening and correcting Liver Qi and Heart.

DOSAGES:

TINCTURE: 1:5, 25% alcohol, 2-4 ml three times daily.

INFUSION: 1:10, 25-50 ml three times daily or 1/2-1 tsp in 1 cup of water, drunk three times daily. Crushed plant and poultices for external use.

- Smaller doses are required for children.
- This herb was formerly called "liver plant" in Icelandic.

GREAT BURNET

BLOODWORT

RANGE AND HABITAT: Found widely in grasslands of South-west Iceland, but rare elsewhere.

PARTS USED MEDICINALLY: The whole plant, with the exception of the root.

GATHERING: All summer long.

ACTIVE SUBSTANCES: Tannins, volatile oils, glycosides, flavonoids and vitamin C.

MEDICINAL ACTION: Astringent and vulnerary.

USES: Burnet is most commonly used for gastritis and pains in the stomach and digestive tract. The herb is recommended for haemorrhoids and persistent diarrhoea. Burnet is also effective as a gargle for sore mouth and a rinse to remedy vaginal discharges (leukorrhea). Like other astringent herbs, burnet can be used on wounds to stop minor bleeding. Dried crushed burnet can therefore be inhaled as snuff to stop nosebleeds.

CHINESE MEDICINE: Chinese herbalists use burnet root, which is acidic and cold and has a strengthening effect on the Liver, Colon and Stomach. The root cools the Blood and stops bleeding in Lower Burner (the body below the navel), the symptoms of which are blood in stools and/or heavy menstruation. The powdered root is also used topically on wounds which heal poorly and on burns, as it both cools swelling of the skin and promotes the growth of new skin. Laboratory research has shown the root to have a retarding effect on various disease-causing bacteria.

DOSAGES:

TINCTURE: 1:5, 45% alcohol, 1-5 ml three times daily.
INFUSION: 1:10, 75-100 ml three times daily or 1 tsp in 1 cup of water, drunk three times daily.
Poultices and crushed herb for external use. A mixture of the dried root powder with sesame oil is effective in treating burns and wounds which heal poorly. Research has shown that powder ground from the entire plant and root is most useful in treating second and third degree burns; the powder is placed directly on the burnt area in poultices. Infusion and diluted tincture as mouthwash.

- Smaller doses are required for children.
- Burnet is an ancient medicinal herb which was widely used. Its leaves and shoots, picked before the plant flowers, have also been used in salads and soups.

Hieracium spp. – Daisy family (Asteraceae)

HAWKWEED

RANGE AND HABITAT: Grows in many types of dry soil. Common everywhere in Iceland.

PARTS USED MEDICINALLY: The whole plant, with the exception of the root.

GATHERING: Early summer.

ACTIVE SUBSTANCES: Volatile oils, pilosellin, bitter principles, tannins, coumarin, flavonoids and antibacterial substances.

MEDICINAL ACTION: The fresh herb is thought to have an antibacterial effect. Fresh and dried, it is astringent and diuretic. It strengthens and stimulates the liver and has antiflatulent and expectorant effects.

USES: Hawkweed is considered useful for digestive ailments resulting from liver malfunction, and is also good in combating flatulence, inflammation of the digestive tract, constipation and diarrhoea resulting from insufficient stomach acidity. The herb is thought to remove gallstones if used as part of more extensive treatment. Hawkweed is very useful for oedema, especially caused by liver malfunction. It has also been used extensively for mucous problems in the respiratory system.

Crushed leaves, placed on burns before blisters develop are a good vulnerary.

DOSAGES:

TINCTURE: 1:5, 25% alcohol, 1-5 ml three times daily.
INFUSION: 1:10, 100 ml three times daily or 1 tsp in 1 cup of water, drunk three times daily.
Fresh, crushed leaves are placed on burns and wounds.

- Smaller doses are required for children.
- Hawkweed is a collective name for several dozen closely related species.

Viola tricolor – Violet family (Violaceae)

HEARTSEASE

RANGE AND HABITAT: Common in some areas of the country, but only an alien elsewhere. Grows in dry gravelly slopes and frequently alongside roads.

PARTS USED MEDICINALLY: The dried plant in flower, with the exception of the root.

GATHERING: Early summer.

ACTIVE SUBSTANCES: Salicylic acid, salicylate, alkaloids, flavonoids, including viola-quercetin, rutin, tannins and mucilage.

MEDICINAL ACTION: Anti-inflammatory, diuretic, expectorant, laxative and sudorific.

USES: Heartsease is used as a blood-purging medicine for all sorts of skin conditions. It is considered especially good for auto-immune illnesses and children's allergy rashes. It is also recommended for all sorts of rheumatism and lung illnesses.

CHINESE MEDICINE: Heartsease is strengthening for Liver Yin and Blood.

DOSAGES:

TINCTURE: 1:5, 25% alcohol, 1-5 ml three times daily.
INFUSION: 1:10, 100 ml three times daily or 1 tsp in 1 cup of water, drunk three times daily.

- Smaller doses are required for children.

Dactylorhiza maculata – Orchid family (Orchidaceae)

HEATHLAND SPOTTED ORCHID

RANGE AND HABITAT: Grows in moorland and brushland. Found widely throughout most of Iceland.

PARTS USED MEDICINALLY: Root.

GATHERING: Late summer and autumn.

ACTIVE SUBSTANCES: Resins and sugars.

MEDICINAL ACTION/USAGE: The root was considered to increase fertility in both men and women and to prevent miscarriage early in pregnancy. It was also thought to strengthen the womb and prepare the body for giving birth. Heathland spotted orchid was considered very effective against cough and hoarseness.

A liniment made from the roots is considered effective against rash and difficult sores.

DOSAGE:

DECOCTION: 1:10, 50 ml three times daily or 1 tsp in 1 cup of water, drunk three times daily. Liniment for external use.

- The plant is not recommended for children.
- The traditional belief in orchids as an aphrodisiac can most likely be attributed to its root, which can resemble male genitalia.

Cetraria islandica – lichen family (Parmeliaceae)

ICELAND MOSS

RANGE AND HABITAT: Common throughout most of Iceland, especially on heaths and mountain highlands. Grows in moorlands.

PARTS USED MEDICINALLY: Entire plant.

GATHERING: Early summer. Iceland moss is thought to be best gathered from wet soil.

ACTIVE SUBSTANCES: Mucilages, including lichenin, bitter acids, minerals, e.g. both iron and potassium salts, and antibacterial substances.

MEDICINAL ACTION: Demulcent and vulnerary for the digestive tract and respiratory organs, also used for external healing. Iceland moss is also very nutritious and therefore a very healthy food.

USES: Iceland moss is among the best herbs for ulcers and inflammation of the digestive tract. It reduces irritation from stomach acids, forming a soothing film over mucous membrane. Iceland moss has long been used for respiratory ailments, especially for persons plagued by a dry cough.

Poultices of Iceland moss are used to treat painful and dry skin eczema.

CHINESE MEDICINE: Iceland moss strengthens Stomach and Lung Yin. It is cooling and nutritive and is used to treat inflammation and pain in the stomach and respiratory tract. Because of its mucilages, Iceland moss cannot be used when the body has excess mucous.

DOSAGES:
TINCTURE: 1:5, 25% alcohol, 2-5 ml three times daily.
INFUSION or decoction: 1:10, 100 ml three times daily or 1 tbsp in 1 cup of water, drunk three times daily. The taste of Iceland moss is not as bitter if boiled in milk.
Poultices for external use.

- Smaller doses are required for children.
- In earlier times, Iceland moss was used as a dye, to produce both yellow and red colours. The yellow colour was obtained by boiling the wool (the garment) with alum as a fixer (as well as Iceland moss). The material acquired a red colour if placed in cow's urine and the liquid changed at two-day intervals.

Cardamine nymanii – Mustard family (Brassicaceae)

LADY SMOCK

BITTERCRESS

RANGE AND HABITAT: Very common throughout Iceland. Grows in fens and marshlands.

PARTS USED MEDICINALLY: The whole plant, with the exception of the root.

GATHERING: Early summer.

ACTIVE SUBSTANCES: The plant has not been researched extensively, but is known to contain both glycosides and bitter principles.

MEDICINAL ACTION: Purges the blood, stimulates menstrual bleeding, kills intestinal parasites, stimulates digestion and liver functioning and strengthens the lungs.

USES: Bittercress has mostly been used to stimulate menstruation. The herb is very bitter and therefore useful to stimulate appetite if taken half an hour before a meal.

Bittercress infusion is very strengthening for weakened persons who have been struggling with a long illness.

CHINESE MEDICINE: Bittercress strengthens Liver and Spleen.

DOSAGES:

TINCTURE: 1:5, 45% alcohol, ½-2 ml three times daily.
INFUSION: 1:10, 25–50 ml three times daily or ½ tsp in 1 cup of water, drunk three times daily.

- In his Travel Journal, the 18th-century natural scientist Sveinn Pálsson wrote that the plant provided good protection against scurvy and had a pleasantly acrid taste.

WARNING!
Pregnant women should completely avoid bittercress; it could cause a miscarriage, especially during the first months of pregnancy. The same applies to most other bitter herbs. The plant is not recommended for children.

Galium verum – Madder family (Rubiaceae)

LADY'S BEDSTRAW

RANGE AND HABITAT: Common throughout Iceland. A sun-loving plant which grows in many types of dry soils.

PARTS USED MEDICINALLY: The whole plant as soon as it flowers, with the exception of the root.

GATHERING: Midsummer.

ACTIVE SUBSTANCES: Silicic acid, glycosides, tannins, volatile oils, plant acids, the enzyme rennin and vitamin C.

MEDICINAL ACTION: Diuretic, purges the blood, vulnerary and astringent, relieves urinary cramps.

USES: Lady's bedstraw is especially useful to treat all types of skin ailments, such as eczema, psoriasis and furuncles, when it is most often used together with other herbs.

An infusion of lady's bedstraw is considered useful to purge the blood after using strong drugs, alcohol or immoderate caffeine consumption. Lady's bedstraw can also be used for kidney and bladder ailments.

CHINESE MEDICINE: Lady's bedstraw is warming and eliminates Dampness and Phlegm. It has a strengthening effect on Liver and Kidney Qi.

DOSAGES:

TINCTURE: 1:5, 25% alcohol, 1-5 ml three times daily.
INFUSION: 1:10, 100 ml three times daily or 1-2 tsp in 1 cup of water, drunk three times daily.

- Smaller doses are required for children.

Alchemilla vulgaris – Rose family (Rosaceae)

LADY'S MANTLE

LION'S FOOT, BEAR'S FOOT

RANGE AND HABITAT: Grows in many types of dry soil. Common throughout Iceland.

PARTS USED MEDICINALLY: Leaves.

GATHERING: Early summer.

ACTIVE SUBSTANCES: Tannins, salicylic acid, phytosterol, saponins, almitin and stearic acid.

MEDICINAL ACTION: Astringent, vulnerary and roborant, stops internal and external bleeding, regulates the menstrual cycle and strengthens the uterus and digestive system.

USES: Lady's mantle is primarily a woman's herb. It relieves menstrual pain and excessive bleeding and is also used to stop breakthrough bleeding. Due to its styptic effects, lady's mantle is used with other herbs to heal ulcers of the stomach and digestive tract.

The plant is considered useful to prevent miscarriage, if an infusion is drunk regularly during the first weeks of pregnancy.

Lady's mantle is a useful douche for excretions and pain in the vaginal tract.

CHINESE MEDICINE: Lady's mantle strengthens the Spleen and Uterus and is used for excessive menstrual bleeding and also to reduce blisters and vaginal and ovarian knots.

DOSAGES:

TINCTURE: 1:5, 25% alcohol, 1-5 ml three times daily.
INFUSION: 1:10, 100 ml three times daily or 1 tsp in 1 cup of water, drunk three times daily.
RINSE: undiluted infusion or tincture, 5 ml in half a cup of water.

- Formerly the herb was used mainly to heal wounds, both externally and in the digestive tract. In ancient times it was dedicated to the goddesses Freyja and Frigg, and later to the Virgin Mary, like many other medicinal herbs.

Vaccinium vitis-idaea – Heather family (Ericaceae)

LINGONBERRY

COWBERRY

RANGE AND HABITAT: Found in woodlands of the East Fjords and in Northeast Iceland, very rare.

PARTS USED MEDICINALLY: Leaves and berries.

GATHERING: Spring and early summer (leaves) and autumn (berries).

ACTIVE SUBSTANCES: Sugars, ericholine, arbutin, tannins and other organic acids.

MEDICINAL ACTION: The leaves are diuretic, antibacterial and astringent, as well as eliminating stones in the urinary tract. The berries are astringent and cooling.

USES: Lingonberries are useful to treat all infections and stones in the urinary organs. They are also recommended for all types of rheumatism.

The berries, which are rich in vitamin C, are considered tasty, stimulate the appetite and are useful to treat diarrhoea.

CHINESE MEDICINE: Lingonberry leaves are warming and strengthening for Kidney Yang and Qi. For this reason they are used for arthritis in the elderly which is helped by heat.

DOSAGES:

TINCTURE: 1:5, 45% alcohol, 1-5 ml three times daily.
INFUSION: 1:10, 100 ml three times daily or 1 tsp in 1 cup of water, drunk three times daily.

- Smaller doses are required for children.
- Because of its rarity in Iceland, it is recommended that people cultivate this plant for their own use.

Caltha palustris – Buttercup family (Ranunculaceae)

MARSH MARIGOLD

RANGE AND HABITAT: Common throughout lowland Iceland. Grows in wetlands and beside streams and ditches.

PARTS USED MEDICINALLY: Dried leaves and flowers.

GATHERING: Early summer.

ACTIVE SUBSTANCES: Protoanemonin, flavonoids, carotene, tannins and saponins.

MEDICINAL ACTION: Analgesic, sudorific, antispasmodic, expectorant, irritates the skin.

USES: Marsh marigold has long been used to treat warts by rubbing fresh flowers on the warts.

Fresh leaves were also placed on sores to clean and heal them.

Marsh marigold has a long history as a treatment for spasms in both children and adults. It can also be used to relieve strong menstrual cramps.

CHINESE MEDICINE: Marsh marigold corrects Liver energy and relaxes.

DOSAGES:

TINCTURE: 1:5, 25% alcohol, 1-3 ml three times daily.

INFUSION: 1:10, 25–50 ml three times daily or ½ tsp in 1 cup of water, drunk three times daily. The fresh herb for external use.

- Smaller doses are required for children.
- The flower buds are considered good to eat. When marsh marigold leaves fell, it was a sign to farmers to begin hay making.

WARNING!

Fresh marsh marigold is a strong irritant and can cause toxicity. For this reason it must always be well dried or boiled before using as medicine. Drying and boiling destroy the substances which cause irritation (protoanemonins).

Ranunculus acris – buttercup family (Ranunculaceae)

MEADOW BUTTERCUP

RANGE AND HABITAT: Common throughout Iceland. Grows in grasslands.

PARTS USED MEDICINALLY: The fresh herb.

GATHERING: Before flowering.

ACTIVE SUBSTANCES: Animonol (highly irritating to the skin).

MEDICINAL ACTION: Irritating, relieves pain and spasms, causes skin reddening.

USES: Meadow buttercup is toxic and cannot therefore be used internally.

It is used externally in poultices for rheumatism, especially gout. Poultices of the fresh plant have also been thought to alleviate pain in the back and joints and headaches.

DOSAGES:

Poultices for external use.

- A traditional remedy was to boil meadow buttercup in milk from an almost dry cow as a drink for the sick.

WARNING!

Always seek the advice of an herbalist before using meadow buttercup. The plant is highly irritating and will raise blisters on the skin if allowed to remain too long. Avoid allowing the poultices to remain longer than just until a redness can be detected.

Filipendula ulmaria – Rose family (Rosaceae)

MEADOWSWEET

MEAD SWEET

RANGE AND HABITAT: Grows in marshlands, woods and fens. Common in the south and southwest of Iceland, rarer in other parts of the country.

PARTS USED MEDICINALLY: Leaves and flowers.

GATHERING: Early summer.

ACTIVE SUBSTANCES: Tannins, volatile oils containing, for instance, salicylaldehyde and salicyl, also glycosides, including salisin and golterin, and mucilage and the flavone glycoside spiracoside.

MEDICINAL ACTION: Demulcent and vulnerary for the stomach lining, reduces secretion of stomach acid, anti-inflammatory, diuretic, astringent and antifebrile.

USES: Meadowsweet has been called both the "stomach's friend" and the herbalist's aspirin. The plant is especially useful to treat stomach ulcers and gastritis, and also reduces high stomach acidity. Because of its salicyl glycosides meadowsweet is good for all sorts of rheumatism and swollen joints, muscles and nerves.

Meadowsweet is astringent and therefore counteracts loose stools and bleeding in the digestive tract. It is used for persisted diarrhoea, haemorrhaging in the digestive tract and haemorrhoids.

Meadowsweet is a very mild herb and has been used extensively for children, both to reduce fever and also for diarrhoea and digestive tract inflammation.

CHINESE MEDICINE: Meadowsweet is cooling and astringent and strengthens the Spleen, Stomach, Small Intestine and Colon. It eliminates Heat from the digestive tract and joints.

DOSAGES:

TINCTURE: 1:5, 25% alcohol, 4-6 ml three times daily.

INFUSION: 1:10, 150 ml three times daily or 1-2 tsp in 1 cup of water, drunk three times daily.

- Smaller doses are required for children.
- Flowers and leaves of meadowsweet were formerly used to flavour ale and wine, as its alternate name, meadsweet, indicates.

Dryas octopetala – Rose family (Rosaceae)

MOUNTAIN AVENS

RANGE AND HABITAT: Grows in dry gravelly and moorland soils, very common throughout Iceland.

PARTS USED MEDICINALLY: The whole plant, with the exception of the root.

GATHERING: Early summer.

ACTIVE SUBSTANCES: Tannins, silicic acid and various minerals.

MEDICINAL ACTION: Astringent, strengthening and slightly stimulating for the digestion.

USES: Mountain avens can be used for ulcers of the stomach and digestive tract, especially where haemorrhaging occurs. The plant is also useful to temper bowel movements and immoderate mucous excretion of the digestive organs. Mountain avens is also thought to strengthen a weak heart.

It is useful as a mouthwash for inflammation and ulcers in the gums, mouth and throat. Infusion of mountain avens is a useful douche for excretions and pain in the vaginal tract.

CHINESE MEDICINE: Mountain avens strengthens Spleen Qi.

DOSAGES:

TINCTURE: 1:5, 25% alcohol, 1-5 ml three times daily.
INFUSION: 1:10, 100 ml three times daily or 1 tsp in 1 cup of water, drunk three times daily. Infusion and diluted tincture as mouthwash.

- Smaller doses are required for children.
- In former times, leaves of mountain avens were often used as herb tobacco, i.e. mixed with smoking tobacco.

Rumex longifolius – Buckwheat family (Polygonaceae)

NORTHERN DOCK

RANGE AND HABITAT: A common naturalised plant in cultivated areas and neglected gardens, generally considered a weed.

PARTS USED MEDICINALLY: Roots, leaves and seeds.

GATHERING: All summer long.

ACTIVE SUBSTANCES: Oxalic acid and other organic acids, e.g. tannic acid, also sugars and vitamins.

MEDICINAL ACTION: Diuretic, laxative, astringent and purges the blood.

USES: The entire plant, but especially the root, is considered very useful to treat all sorts of dermal ailments, especially those accompanied by severe itching. The plant is used both internally and externally. The entire plant is useful to treat burns, sores which are slow to heal and skin inflammation or infections.

In other respects, northern dock has similar effects to common sorrel.

WARNING!
Because of the high acidic content of northern dock, persons suffering from arthritis and excess stomach acid should not use the plant internally.

DOSAGES:
TINCTURE: 1:5, 45% alcohol, 2-4 ml three times daily.
INFUSION of seeds and leaves and decoction of roots: 1:10, 40-60 ml three times daily or 1/2-1 tsp in 1 cup of water, drunk three times daily.
Strong infusion of roots for external use.
Liniment for external use.

- Smaller doses are required for children.
- Fresh leaves are healthy greens (were referred to as "flitting days' kale") and can also be used in soups, porridge and cream sauces. As a folk remedy for headache, root of northern dock was split and the two halves placed one on each side of the head, with the cut side inward.

Trifolium pratense – Pea family (Fabaceae)

RED CLOVER

TREFOIL OR PURPLE CLOVER

RANGE AND HABITAT: A naturalised plant in grasslands throughout much of Iceland.

PARTS USED MEDICINALLY: Flowers.

GATHERING: Early summer.

ACTIVE SUBSTANCES: Salicylic acid and other organic acids, phenolic glycosides and other glycosides, sugars, flavonoids, coumarin and phytoestrogens.

MEDICINAL ACTION: Vulnerary, stimulant, diuretic and expectorant.

USES: Red clover is among the best herbs to treat all types of eczema and psoriasis, not least in children. Red clover has been used for cancer, especially breast and ovarian cancer.

A liniment prepared from the flowers has been used extensively to treat skin cancer (see the recipe for tar liniment in the section on preparing herbal medicines).

Because of the phytoestrogens in the flowers, they have been used to enhance female fertility and also to treat menopause symptoms, especially nocturnal sweating and difficulty sleeping.

CHINESE MEDICINE: Red clover flowers strengthen Kidney Yin and are used for eczema characterised by Dryness. They strengthen Bones and Marrow and are used to treat osteoporosis and rheumatism in the elderly.

DOSAGES:

TINCTURE: 1:5, 25% alcohol, 1-5 ml three times daily.
INFUSION: 1:10, 100 ml three times daily or 1-2 tsp in 1 cup of water, drunk three times daily.
Liniment for external use.

- Smaller doses are required for children.
- Young leaves, picked before the plant flowers, are good in salads and soups.

Plantago lanceolata – Plantain family (Plantaginaceae)

RIBWORT PLANTAIN

ENGLISH PLANTAIN, BUCKHORN PLANTAIN, NARROWLEAF PLANTAIN (MANY OTHER NAMES)

RANGE AND HABITAT: Rare except in the extreme south of the country. Grows on grassy slopes and geothermal areas.

PARTS USED MEDICINALLY: Leaves.

GATHERING: Early summer.

ACTIVE SUBSTANCES: Mucilage, glycosides, including aucubin, tannins, silica compounds and various minerals.

MEDICINAL ACTION: Strengthens mucous membrane in the respiratory tract, helps stop runny nose, demulcent and cooling, both internally and externally.

USES: Ribwort plantain is used extensively to strengthen the mucous membrane in the throat, ears and sinuses. It is considered useful in treating colds.

Ribwort plantain is also considered useful for spasmodic coughing, as the herb is both strengthening and demulcent.

The herb can be used externally for inflammation of the skin.

CHINESE MEDICINE: Chinese physicians use both the leaves and seeds but for different purposes. Ribwort plantain seeds are sweet and cold and have a strengthening effect on the Bladder, Kidneys, Liver and Lungs. They are diuretic and release Heat from the body. This makes them useful against inflammation and pain in the bladder and urinary tract caused by Damp Heat. They clear the eyes and are used for eye ailments caused by an imbalance of the Liver and Kidneys, especially where Heat is involved. They resolve Dampness and are used to treat coughs caused by Damp Heat in Lungs.

The leaves are sweet and cold and considered useful in releasing Heat from the body and reducing Heat toxicity. The leaves are thus used for furuncles and inflamed lumps.

DOSAGES:

TINCTURE: 1:5, 45% alcohol, 1-5 ml three times daily.
INFUSION: 1:10, 75-100 ml three times daily or 1 tsp in 1 cup of water, drunk three times daily.
Poultices and liniment for external use.

- Smaller doses are required for children.
- Fresh ribwort plantain leaves are good in salads and soups and the mature seeds are highly recommended as birdseed.

Rhodiola rosea – Crassulaceae

ROSEROOT

GOLDEN ROOT

RANGE AND HABITAT: Fairly common throughout most of Iceland. Grows mainly on rock and other places less frequented by sheep.

PARTS USED MEDICINALLY: Rhizome.

GATHERING: Late summer and autumn.

ACTIVE SUBSTANCES: Tannins, but little research has been done on the plant's active substances.

MEDICINAL ACTION: Astringent, vulnerary, anti-inflammatory and also cleans sand from the kidneys.

USES: Roseroot has been used primarily for its astringent qualities. It is considered effective against inflammation and pain in the digestive tract and was also used for diarrhoea and bacillary dysentery.

As a mouthwash it can be effective against inflammation and pain in the mouth, and as a douche for pain and discharges from the vagina. Poultices of crushed rhizomes are used for skin lesions and inflammation.

CHINESE MEDICINE: Roseroot strengthens Spleen Qi.

DOSAGES:

TINCTURE: 1:5, 25% alcohol, 1-3 ml three times daily.
DECOCTION: 1:10, 25-50 ml three times daily or 1/2-1 tsp in 1 cup of water, drunk three times daily.
Infusion and tincture as mouthwash.
Poultices and liniment for external use.

- In earlier times, lepers were advised to used this plant as treatment. An infusion of the plant was thought to stimulate hair growth if used on the hair in the morning and evening to dampen the scalp. This had to be done for a lengthy period, however.

Sorbus aucuparia – Rose family (Rosaceae)

ROWAN

EUROPEAN ROWAN, MOUNTAIN ASH

RANGE AND HABITAT: Various places around the country, especially in birch woods and ravines. It is also a common garden tree.

PARTS USED MEDICINALLY: Berries.

GATHERING: Autumn.

ACTIVE SUBSTANCES: Tannins, sorbitol, ascorbic acid, malic acid, vitamin C and sugars.

MEDICINAL ACTION: Astringent, laxative, diuretic, stimulates menstruation.

USES: Fresh berry juice is used as a laxative, especially for children. It is used for ulcers of the lining of the digestive tract, sore throats and colds and even tuberculosis. The juice is also used to stimulate menstruation.

Dried or stewed berries have an astringent effect and are used to treat diarrhoea, especially in children.

Berries are used with other herbs to treat oedema.

DOSAGES:

TINCTURE: 1:5, 25% alcohol, 1-4 ml three times daily.
INFUSION: 1:10, 50-75 ml three times daily or 1 tsp in 1 cup of water, drunk three times daily.
Fresh juice, 5-10 ml three to four times daily.

- Smaller doses are required for children.
- In heathen times, the rowan was dedicated to the god Thor and called "Thor's rescuer", as in mythology the god Thor managed to survive an encounter with a giantess by grasping a rowan branch to escape a river in flood. In Christian times the tree was much revered, and was not to be cut down or pruned.

Ligusticum scoticum – Umbelliferae

SCOTS LOVAGE

RANGE AND HABITAT: Found on ocean cliffs and talus. A rare plant, found especially in West Iceland.

PARTS USED MEDICINALLY: Roots and seeds.

GATHERING: The root in early summer, the seeds in autumn.

ACTIVE SUBSTANCES: Volatile oils, mannitol, glycosides, vitamins and various minerals.

MEDICINAL ACTION: Analgesic and antiflatulent, stimulates digestion, expectorant, sudorific and stimulates menstruation.

USES: Scots lovage has effects similar to those of other angelica species. The root stimulates appetite and reduces flatulence. Because of its oils, Scots lovage is also useful to treat infections of the lungs and urinary organs.
The seeds are sedative and good for stress.

CHINESE MEDICINE: Scots lovage is bitter and warming and affects the Liver and Gallbladder and Pericardeum. It corrects Liver Qi and eliminates Dampness and Phlegm. It loosens pent up energy and is useful to treat pressure and pain in the chest and pelvic girdle. Roots of Scots lovage are important gynecologically, as they stimulate menstruation and relieve menstrual pain. The herb is also used to ease giving birth.

DOSAGES:

TINCTURE: 1:5, 45% alcohol, 2-3 ml three times daily.
INFUSION of seeds and decoction of roots: 1:10, 25-50 ml three times daily or 1/2 tsp in 1 cup of water, drunk three times daily.

- Smaller doses are required for children.
- Because of its rarity in Iceland, it is recommended that people cultivate this plant for their own use.

Plantago maritima – Plantain family (Plantaginaceae)

SEA PLANTAIN

RANGE AND HABITAT: Common throughout Iceland. Grows in deserted farmland, in rocky and gravelly areas, coastal vegetation patches and river banks.

PARTS USED MEDICINALLY: Leaves and roots.

GATHERING: Early summer prior to blooming.

ACTIVE SUBSTANCES: Tannins, mucilage, silicic acid, aucubin and vitamin C.

MEDICINAL ACTION: The plant is regarded as roborant, vulnerary, astringent, demulcent and anthelmintic.

USES: Sea plantain was used for lack of appetite, inflammation and pain in the digestive tract and diarrhoea.
The plant was also used in a manner similar to ribwort plantain, to treat colds and sinusitis.

DOSAGES:
TINCTURE: 1:5, 45% alcohol, 2-4 ml three times daily.
INFUSION of leaves and decoction of roots: 1:10, 50 ml drunk three times daily or 1 tsp in 1 cup of water, drunk three times daily.

- Sea plantain was eaten in times of famine and was thought to eliminate intestinal parasites.

Prunella vulgaris – Bergamot family (Lamiaceae)

SELFHEAL

PRUNELLA, ALL-HEAL, HEART OF THE EARTH

RANGE AND HABITAT: Common in the warmer regions of Iceland. Grows in grassland and brush.

PARTS USED MEDICINALLY: The entire plant in flower, with the exception of the root.

GATHERING: All summer long.

ACTIVE SUBSTANCES: Bitter principles, tannins, volatile oils and sugars, vitamins K, B1 and C.

MEDICINAL ACTION Vulnerary, roborant, antispasmodic and astringent.

USES: Selfheal is especially effective in healing lesions, both to the skin and in the digestive tract. The plant is recommended to staunch bleeding in the digestive tract and for diarrhoea. Selfheal has been used successfully for minor cramps and convulsions.

Gargling with tea made from the leaves can relieve sore throats.

A poultice of the plant can be effective if applied to wounds which are slow to heal.

Selfheal is a mild herb which can therefore be used by children and adults alike.

CHINESE MEDICINE: Selfheal is cooling and has the strongest impact on the Gall bladder and Liver. It resolves Liver Heat/Fire and clears eyes. It is used to treat conjunctivitis and bloodshot eyes, as well as for eye pain that worsens in the evenings. Selfheal is also used for hypertension, headaches and fainting caused by imbalance of Liver energy. Selfheal rids the body of heat and loosens knots (damming energy and substances) and for this reason is used for swollen glands.

DOSAGES:

TINCTURE: 1:5, 25% alcohol, 1-5 ml three times daily.

INFUSION: 1:10, 100 ml three times daily or 1 tsp in 1 cup of water, drunk three times daily.

- Smaller doses are required for children.
- Selfheal can be used with other herbs in salads and soups.

Capsella bursa-pastoris – Mustard family (Brassicaceae)

SHEPHERD'S PURSE

RANGE AND HABITAT: Common in lowland areas throughout Iceland, especially neglected gardens.

PARTS USED MEDICINALLY: Entire plant except the roots.

GATHERING: All summer long.

ACTIVE SUBSTANCES: Choline, acetylcholine, tyramine, amino acids, sugars, flavonoids, various minerals and vitamins, especially vitamins A, B and C.

MEDICINAL ACTION: Stops bleeding both internally and externally, causes contraction of blood vessels, diuretic and strengthens and disinfects the mucous membrane of urinary organs.

USES: Infusions and tinctures of shepherd's purse have been used extensively with good results to treat excessive menstruation. If the plant is taken for some time it can stop breakthrough bleeding in the menstrual cycle. It was also used to stop post-partum haemorrhaging. If Shepherd's purse is to reduce menstrual bleeding during the normal period of the cycle, it must be used from the midpoint of the cycle until the bleeding stops. This is repeated during the next cycle. Shepherd's purse is considered useful for various types of urinary infections, especially when there is much mucous discharge or bleeding from the urinary tract.

A powder prepared from the plant is useful for stopping nose bleeds, as well as minor bleeding from dermal sores.

CHINESE MEDICINE: Shepherd's purse strengthens the Spleen and the Uterus and eliminates Dampness and Plegm, especially in the pelvic area.

DOSAGES:

TINCTURE: 1:5, 25% alcohol, 5-7 ml three times daily.
INFUSION: 1:10, 100–200 ml three times daily or 1-2 tsp in 1 cup of water, drunk three times daily. Powder for external use and in the nostrils. Fresh, crushed plant for external use.

- Smaller doses are required for children.
- In cases of internal haemorrhaging, the advice of a physician should always be sought before attempting to treat the ailment with herbs.
- In earlier times, shepherd's purse was used against malaria if quinine was not available.

Urtica dioeca – Nettle family (Urticanceae)

STINGING NETTLE

RANGE AND HABITAT: Rare, non-indigenous species in Iceland. Grows in settled areas, usually in or near gardens.

PARTS USED MEDICINALLY: Leaves, which are only used dried.

GATHERING: Early summer.

ACTIVE SUBSTANCES: Alkaloids, which contain for instance formic acid and acetic acid, also histamines, tannins, vitamins A and C and many minerals, e.g. iron.

MEDICINAL ACTION: Astringent, stops both internal and external bleeding, nourishing, diuretic, stimulates lactation in nursing mothers and lowers blood sugar.

USES: Nettles are very useful to treat rashes and eczema, especially blister rashes which are very itchy.

The herb is very useful for children prone to allergy rashes. Stinging nettle is nutritive and therefore often recommended for people suffering from anaemia or malnourished.

The herb can be used for all internal haemorrhaging and heavy menstruation.

Stinging nettle is used to stimulate lactation in nursing mothers and due to its effect in lowering blood sugar is useful to treat Type 2 diabetes.

CHINESE MEDICINE: Stinging nettle cools the Blood and strengthens Liver Yin. Because of this it is used for all allergies, even auto-immune disorders. It quietens the Spirit and reduces itching and irritation in the body.

DOSAGES:

TINCTURE: 1:5, 45% alcohol, 2-5 ml three times daily.
INFUSION: 1:10, 75-100 ml three times daily or 1 tsp in 1 cup of water, drunk three times daily.

- Smaller doses are required for children.
- Lesser nettle or Small nettle (*Urtica urens*) has an effect similar to stinging nettle and can therefore be used in the same manner.
- Boiled new tips can be used for making soup.

WARNING!
Fresh nettle leaves cause irritation or rash, both externally and internally.

Rubus saxatilis – Rose family (Rosaceae)

STONE BRAMBLE

RANGE AND HABITAT: Fairly common almost everywhere in Iceland. Grows in grassland and woodlands.

PARTS USED MEDICINALLY: Roots, leaves and berries.

GATHERING: All summer long.

ACTIVE SUBSTANCES: Tannins, sugars, acids and vitamin C.

MEDICINAL ACTION: Astringent, purges the blood, strengthen-ingand vulnerary.

USES: Stone bramble is used for pain and inflammation of the digestive tract and also for diarrhoea and mucous stools. Dried berries, ground to a powder, can be used in tea, which is considered very effective to treat diarrhoea in children. All parts of the plant are considered strengthening and are very suitable after chronic illness or operations.

Stone bramble can be used as a gargle for pain and inflammation in the mouth and a rinse to remedy vaginal discharges (leukorrhea).

CHINESE MEDICINE: Stone bramble strengthens Spleen and Stomach.

DOSAGES:

TINCTURE: 1:5, 25% alcohol, 1-5 ml three times daily.
INFUSION of leaves and berries and decoction of roots: 1:10, 100 ml three times daily or 1 tsp in 1 cup of water, drunk three times daily.
Powder of berries for internal use.

- Smaller doses are required for children.
- The berries are good to eat fresh and as jam. The plant's runners were formerly called "devil's rope" or "trolls' rope" and according to folklore, could be used to fetter evil spirits.

Myrrhis odorata – Apiaceae

SWEET CICELY

BRITISH MYRRH, ANISE, SWEET CHERVIL

RANGE AND HABITAT: A naturalised plant in gardens and near farmhouses.

PARTS USED MEDICINALLY: The entire plant, including the seeds.

GATHERING: The leaves are best in early summer, while other parts are better later.

ACTIVE SUBSTANCES: Volatile oils containing, for instance, anethole, bitter principles, various other oils and glycosides, including glycyrrhizin.

MEDICINAL ACTION: Stimulates the digestion, antispasmodic, analgesic and antiflatulent, expectorant, relieves hypertension somewhat and strengthens the hormone system.

USES: Sweet cicely is mostly used to improve digestion. It is considered useful to relieve pain caused by flatulence and intestinal cramps. The entire plant is useful as cough medicine, as it strengthens the lungs and expels persistent phlegm. The root is useful for youths and women during menopause. The root is thought to be antiseptic and especially useful to treat all sorts of respiratory infections.

The entire plant is very mild and can therefore be used by children and the elderly alike.

CHINESE MEDICINE: Sweet cicely strengthens the Liver, Lungs and Spleen and corrects Liver Qi. It is cooling and eliminates Dampness from the digestive system and Lungs.

DOSAGES:

TINCTURE: 1:5, 45% alcohol, 1-5 ml three times daily.

INFUSION of leaves and seeds and decoction of roots: 1:10, 50 ml three times daily or 1 tsp in 1 cup of water, drunk three times daily.

- Smaller doses are required for children.
- Young roots are good in salads and with various vegetables. They are boiled and diced. The seeds and leaves can also be used to add spice to salads and chopped leaves are good with various types of pickled foods.

Hierochloë odorata – Grass family (Poaceae)

SWEET GRASS

BUFFALO GRASS, HOLY GRASS, MANNA GRASS

RANGE AND HABITAT: Grows in fertile grasslands throughout much of Iceland.

PARTS USED MEDICINALLY: The leaves before flowering.

GATHERING: Early summer.

ACTIVE SUBSTANCES: Not researched.

MEDICINAL ACTION/USAGE: Sweet grass was considered to strengthen the heart, purge the blood and to be diuretic and sudorific. It was used to treat oedema and eczema.

A liniment made from the leaves is useful to treat skin rashes but apart from that sweet grass is little used for medicinal purposes nowadays.

DOSAGES:

Liniment for external use.

- The plant is not recommended for children.
- After drying, the plant has a sweet smell. It was therefore used in clothes chests, where it also repels moths and other insects.

Anthoxanthum odoratum – Grass family (Poaceae)

SWEET VERNAL GRASS

RANGE AND HABITAT: Grows in dry meadows, on slopes and moorlands. Common everywhere in Iceland.

PARTS USED MEDICINALLY: Flowers.

GATHERING: Early summer.

ACTIVE SUBSTANCES: Volatile oils and coumarin.

MEDICINAL ACTION: Sweet vernal grass causes hay fever and pollen allergies but the flowers are thought to offer relief from the immune system responses.

USES: The significance of sweet vernal grass as a medicinal herb is primarily in its effect against pollen allergies. A tincture is prepared in an untraditional manner. It is made from sweet white wine, fortified with distilled spirit to give a strength of 25%. The proportions are 1:5 as usual. When an allergy attack begins, the tincture shall be inhaled well, and 3-5 drips placed on the tongue and swallowed. This is repeated at 30-60 minute intervals while the pollen allergy persists.

- Smaller doses are required for children.
- The leaves give off a sweet scent when dried and were formerly used in clothes chests like vanilla grass.

VALERIAN

RANGE AND HABITAT: Rare in the wild, but common in gardens. Grows in wildflower patches and brushland.

PARTS USED MEDICINALLY: Rhizomes.

GATHERING: In spring or autumn.

ACTIVE SUBSTANCES: Volatile oils, containing for instance, valerianic acid, borneol, pinene and camphene, as well as volatile bitters e.g. catanin, skytantin and resins The fresh root also contains strongly sedative substances.

MEDICINAL ACTION: Sedative, without reducing concentration, diuretic, expectorant and relieves cramps.

USES: Valerian is used extensively for all types of stress and is considered especially good for digestive illnesses caused by stress. The root, especially fresh, is used to treat certain types of insomnia, and in combination with other herbs to combat hypertension. Valerian soothes most organs and is useful in the treatment of tachycardia and other stresses that affect the heart and muscles.

CHINESE MEDICINE: Valerian affects Liver Qi and treats energy blockages in the body. It protects Stomach, Spleen and Lungs against the effects of stress and is therefore a good addition to herbal mixtures for digestive and respiratory illnesses arising from stress.

DOSAGES:

DRIED ROOTS:

TINCTURE: 1:5, 25% alcohol, ½–1 ml three times daily.

INFUSION: 1:10, 10-20 ml three times daily or 1/2-1 tsp in 1 cup of water, drunk three times daily.

FRESH ROOTS:

TINCTURE: 1:2, 45% alcohol, ½–1 ml three times daily.

INFUSION: 1:5, 50 ml in early evening for insomnia or ½–1 tsp: 1 cup of water, drunk three times daily.

WARNING!

If taken in excessive dosages, valerian can cause all the symptoms it is to prevent, e.g. tachycardia and stress. Do not use it together with other hypnotics because it amplifies their effect. The plant is not recommended for children.

Geum rivale – Rose family (Rosaceae)

WATER AVENS

RANGE AND HABITAT: Common practically everywhere in Iceland. Prefers moist soils, grassy moorlands and hollows.

PARTS USED MEDICINALLY: The flowering plant and stolons (runners).

GATHERING: Late summer and autumn.

ACTIVE SUBSTANCES: Volatile oils, including eugenol, bitter principles, tannins, glycosides and flavonoids.

MEDICINAL ACTION: Astringent, roborant, stimulates and strengthens the liver and improves digestion, expectorant.

USES: The runners are both tasty and effective against diarrhoea. The entire plant is useful for lack of appetite and poor digestion.

Water avens is very regenerative for people who have been ill for lengthy periods and an infusion of the plant is thought to prevent motion sickness. An infusion of the flowering plant is also used to treat persistent bronchitis and sinus congestion.

CHINESE MEDICINE: Water avens strengthens Spleen and therefore helps to process energy from food and drink. It also eliminates Dampness and Phlegm from the body.

DOSAGES:

TINCTURE: 1:5, 25% alcohol, 2-3 ml three times daily.

AN INFUSION from the aerial parts of the plant or decoction of the runners: 1:10, 25-50 ml three times daily or 1/2 tsp in 1 cup of water, drunk three times daily.

- Smaller doses are required for children.

Veronica anagallis-aquatica – Plantain family (Plantaginaceae)

WATER SPEEDWELL

BLUE SPEEDWELL, BROOK PIMPERNEL

RANGE AND HABITAT: Very rare. Grows in Southwest Iceland by streams and pools.

PARTS USED MEDICINALLY: The fresh or dried plant in flower, with the exception of the root.

GATHERING: Late summer.

ACTIVE SUBSTANCES: The plant has not been researched extensively, but is known to contain both tannins and glycosides.

MEDICINAL ACTION: Stimulant, diuretic, cough suppressant, antispasmodic and antifebrile.

USES: Although little used now, water speedwell was formerly used to stimulate liver functioning, e.g. to treat jaundice.
The herb was also considered useful for haemorrhoids, gastritis, cramps and other ailments of the digestive organs. It was also recommended for apoplexy.

Used externally, common speedwell is vulnerary and demulcent.

DOSAGES:

INFUSION: 1:10, 50 ml three times daily or 1 tsp in 1 cup of water, drunk twice to three times daily.
Poultices for external use.

- Smaller doses are required for children.
- Powder made from the root was considered helpful for the head and vision.

Catabrosa aquatica – Grass family (Poaceae)

WHORL-GRASS

RANGE AND HABITAT: Fairly common almost everywhere in Iceland. Grows in shallow ponds, ditches and springs.

PARTS USED MEDICINALLY: Entire plant.

GATHERING: Early summer.

ACTIVE SUBSTANCES: The active substances have not been fully researched, but its effect is known to be related to volatile oils.

MEDICINAL ACTION: Stimulates digestion and cleansing for the liver, treats gastroenteritis, stimulates the bowels and menstruation. Used externally, whorl-grass is a good vulnerary.

USES: Whorl-grass is used for constipation and lack of appetite. It is considered especially useful for mucousy stools. The herb is used externally to treat wounds which are slow to heal.

An infusion of whorl-grass is also considered useful as an eyewash for sore eyes.

DOSAGES:

TINCTURE: 1:5, 25% alcohol, 2-4 ml three times daily.
INFUSION: 1:10, 50 ml three times daily or ½ tsp in 1 cup of water, drunk three times daily.
The fresh or dried herb for external use.
Eyewash for external use.

- Smaller doses are required for children.

Fragaria vesca – Rose family (Rosaceae)

WILD STRAWBERRY

RANGE AND HABITAT: Found in many areas of Iceland. A sun-loving plant which flourishes in well vegetated, south-facing slopes .

PARTS USED MEDICINALLY: Berries, leaves and roots.

ACTIVE SUBSTANCES: Tannins, mucilage, salts, e.g. potassium, vitamin C, flavonoids and volatile oils.

MEDICINAL ACTION: Astringent, diuretic, strengthening and purges the blood.

USES: All parts of the plant are considered useful to treat severe skin ailments, especially dermal toxicity. The plant can also be used for poultices or taken orally.

The flowers and root are useful to treat intestinal ailments, e.g. diarrhoea, colitis and cramps. They are also good for urinary illnesses such as cystitis, kidney sand and haemorrhaging.

CHINESE MEDICINE: The strawberry plant and berries are cooling and eliminate Heat and Heat toxicity. The plant strengthens the Liver and purges the Blood.

DOSAGES:

The berries can be eaten fresh and can also be dried and used in infusions or decoctions.

LEAVES AND ROOTS:
TINCTURE 1:5, 25% alcohol, 1-5 ml three times daily.
INFUSION of leaves and berries and decoction of roots: 1:10, 100 ml three times daily.
Or 1 tsp: cup of water, drunk three times daily.

- Smaller doses are required for children.
- An 18th century Icelandic herbal claims that strawberries clean plaque from teeth.

Salix spp. – Willow family (Saliceae)

WILLOW

RANGE AND HABITAT: Several willow species, including Tea-leaved willow, Arctic willow and Woolly willow, are found in Iceland and are common throughout the country. They grow in a variety of soil types and often form brush woods.

PARTS USED MEDICINALLY: Bark of branches two years or older.

ACTIVE SUBSTANCES: Salicylates, including salicin, tannins and resins.

MEDICINAL ACTION: Willow has an effect similar to aspirin and other analgesics containing salicylic acid, insofar as it relieves pain and fever and is anti-inflammatory. In addition, willow is astringent, strengthening and a local antiseptic.

USES: Willow is used extensively for all types of rheumatism, auto-immune complaints and inflammation in the body. It was and is used widely because of its antifebrile qualities and has the advantage over aspirin of being completely harmless. In fact, willow is used extensively for inflammation and pain in the stomach and other parts of the digestive tract.

It is very effective for menstrual pain and has been used together with other herbs to treat endometriosis with good results. Willow has been studied extensively and it is now thought to have an effect on cancer tumours, for which a strong tincture must be used.

The herb can be used externally on burns and severe skin sores.

CHINESE MEDICINE: Willow bark is roborant, especially for the Spleen and eliminates Dampness. It is cooling and therefore a useful medicine for rheumatism and other illnesses characterised by Heat or Dampness, i.e. inflamed and hot joints and pain in the digestive and respiratory tract.

DOSAGES:
TINCTURE: 1:5, 25% alcohol, 3-5 ml three times daily or 1:1, 25% alcohol, 3-5 ml three times daily.
DECOCTION: 1:10, 100 ml three times daily or 1-2 tsp in 1 cup of water, drunk three times daily.
Poultices and strong infusion for external use.

- Smaller doses are required for children.
- All willow species are thought to have similar effects.
- Formerly ash of willow bark was used to remove warts, corns and hair from the legs.

Geranium sylvaticum – Cranesbill family (Geraniaceae)

WOOD CRANESBILL

WOODLAND GERANIUM

RANGE AND HABITAT: Common throughout Iceland. Grows in woodlands, ravines and hollows.

PARTS USED MEDICINALLY: Entire plant and root.

GATHERING: Early summer prior to blooming.

ACTIVE SUBSTANCES: Tannins, resins, geranine and acids.

MEDICINAL ACTION: Vulnerary, astringent, promotes regular bowel functioning and anti-inflammatory.

USES: Wood cranesbill is recommended for inflammation and ulcers of the digestive tract; the plant is also indicated for diarrhoea.

Wood cranesbill has long had a reputation as a treatment for various rheumatic complaints, especially gout. The leaves are recommended for external use. They are boiled and placed on lesions which are slow to heal or badly bruised flesh and chronic eczema. An infusion made from the leaves can be used to rinse secretions from the vaginal tract (leukorrhea).

CHINESE MEDICINE: Wood cranesbill is cooling and strengthens Kidney Yin.

DOSAGES:

UTINCTURE: 1:5, 45% alcohol, 2-4 ml three times daily.

AN INFUSION from the aerial parts of the plant or decoction of the root: 1:10, 50–100 ml three times daily or 1/2 tsp in 1 cup of water, drunk three times daily.

- Smaller doses are required for children.
- The plant is also known as dye-grass from the time it was used as dye.

Anthyllis vulneraria – Pea family (Fabaceae)

WOUNDWORT

RANGE AND HABITAT: Woundwort is very rare in Iceland with the exception of the Reykjanes peninsula in the southwest as far inland as Mosfellssveit. It grows in dry grasslands and sandy soil.

PARTS USED MEDICINALLY: Flowering tips.

GATHERING: Late summer.

ACTIVE SUBSTANCES: Saponins, tannins and mucilage.

MEDICINAL ACTION: Astringent, vulnerary, mildly laxative and cough suppressant.

USES: Woundwort is mostly used externally, on sores and minor burns. An infusion of the plant is useful to wash wounds.

An infusion of woundwort is often given to children suffering from constipation. If constipation persists, the advice of a physician or herbalist should be sought.

Woundwort often reduces nausea and vomiting and has proven useful to many women for treating morning sickness. It is also useful for motion sickness (sea-, airplane- and carsickness).

DOSAGES:

TINCTURE: 1:5, 25% alcohol, 1-5 ml three times daily.
INFUSION: 1:10, 100 ml three times daily or 1 tsp in 1 cup of water, drunk in the morning and the evening.
Poultices and infusion for external use.

- Smaller doses are required for children.
- Woundwort is regarded as a very good feed plant and was likely brought to Iceland for such use.

Achillea millefolium – Daisy family (Asteraceae)

YARROW

RANGE AND HABITAT: Grows on dry slopes and flatlands, e.g. common by the roadside and near farmhouses. Common throughout much of Iceland.

PARTS USED MEDICINALLY: Leaves and flowers.

GATHERING: Early summer.

ACTIVE SUBSTANCES: Volatile oils containing, for instance, cineol, azulene, eugenol, thujone, pinene and camphor, also bitter glycosides, cyanosides, salicylate, asparagine, choline, tannins and isovaleric acid.

MEDICINAL ACTION: Astringent and vasodilative, especially for the peripheral vessel system, thereby lowering blood pressure, also sudorific, antispasmodic, sedative and regulates menstruation

Used externally, yarrow is astringent and vulnerary.

USES: Drinking an infusion of yarrow can be helpful in the advent of a cold or flu. It is also useful against all children's illnesses, such as measles, whooping cough, rubella and other such ailments. Yarrow is often used with other herbs to reduce hypertension and also to purge the blood in cases of rheumatoid arthritis and eczema.

Yarrow is good to relieve cramps and uterine pain, to stimulate menstruation and for heat flushes and insomnia during menopause.

Yarrow has long been regarded as among the best vulneraries for persistent sores. A liniment made from the flowers or gauze soaked in a decoction should be placed on the wound. Bathing with yarrow can also be used to treat various skin conditions, such as allergy rashes and eczema.

CHINESE MEDICINE: Yarrow strengthens Liver and Heart Yin. The herb corrects Liver Qi and therefore has an antispasmodic effect throughout the body. Yarrow is also cooling and for this reason useful in the advent of flu or sore throat accompanied by fever and aches.

DOSAGES:

TINCTURE: 1:5, 25% alcohol, 1-5 ml three times daily.

INFUSION: 1:10, 100 ml three times daily or 1 tsp in 1 cup of water, drunk three times daily. Baths, poultices and liniment for external use.

- Smaller doses are required for children.
- Decoction of yarrow was formerly used to bathe the face, as it was believed to remove wrinkles.

Rhinanthus minor – Broomrape family (Orobanchaceae)

YELLOW RATTLE

COCKSCOMB

RANGE AND HABITAT: Grows in many types of dry soils, common throughout Iceland.

PARTS USED MEDICINALLY: The whole plant, with the exception of the root.

GATHERING: Early summer.

ACTIVE SUBSTANCES: Tannins and glycosides.

MEDICINAL ACTION Expectorant, demulcent and astringent.

USES: Yellow rattle is considered helpful for dry coughs and even asthma.

However, the plant has mostly been used as an eyewash for all types of eye ailments, such as conjunctivitis and sensitive eyes, and to clear eye haze.

DOSAGES:

TINCTURE: 1:5, 25% alcohol, 1-5 ml three times daily.
INFUSION: 1:10, 50–100 ml three times daily or 1 tsp in 1 cup of water, drunk three times daily. Weak infusion and tincture (diluted with water) as eyewash.

- Smaller doses are required for children.
- Children formerly used the fruit of yellow rattle to play shopkeeper, and another name for the plant in Icelandic is "money plant".

GATHERING AND DRYING HERBS AND MIXING OF HERBAL MEDICINES

GATHERING HERBS

When gathering wild herbs make sure to always have good photos or illustrations of the plants to compare them with, because it can often be difficult to distinguish between similar plants and there are a number of poisonous plants in Iceland, e.g. herb paris (*Paris quadrifolia*) and male fern (*Dryopteris filix-mas*).

Take care as well never to pick so much in a single location that there is a danger that the plant will be eradicated, because Icelandic flora is very sensitive. Bear in mind that some plants are protected (see the list of protected plants in Iceland on p. 123).

Since it is very easy to cultivate herbs, it is a good practice, to grow those which you want to use for your health care. And you should certainly do so, in the case of rare Icelandic plants.

Never gather herbs where they could be contaminated, e.g. by automobiles or industry.

Gather herbs in locations where they grow by abundantly, because it is evident that the soil there suits them, and the herbs will be stronger.

Once picked, herbs should be placed in a basket or other container where air can circulate freely around them. Upon returning home, they should be hung up to dry or spread out on a suitable underlay, e.g. a net, so that they are well ventilated.

LEAVES

Gather leaves in the morning, after the dew has dried on them. The leaves are most potent just before the plant blooms. Cut them from the plant using a sharp knife or garden shears, because there is a danger of damaging the stem if the leaves are torn off it. Pick only undamaged and healthy-looking leaves.

FLOWERS

The flowers are best picked in the middle of the day and on a dry, sunny day. The flowers contain the greatest amount of active ingredients when they have just fully opened. Place the picked flowers in a dark place as soon as possible after gathering, spreading them out well so that they dry as quickly as possible. Flowers are very sensitive and should be handled carefully after gathering, so always pick undamaged flowers and do not let them lie in bunches any longer than necessary.

THE WHOLE PLANT, WITH THE EXCEPTION OF THE ROOT

If the whole plant is to be used except the root, it is best picked just before flowering, unless the flowers are also to be used. Cut annuals 10 cm above the ground and never take more than one-third of perennials.

SEEDS AND FRUIT

Pick seeds and fruit on a dry day when they are fully mature. The seeds should be golden brown, brown or black but never green when they are picked. Shake a few seeds off into a paper bag, or clip the flower heads off and hang them over a tray to collect the seeds. Mark the packets immediately to avoid any possible confusion concerning the contents.

ROOTS AND STOLONS (RUNNERS)

The roots are most potent in the autumn when the aerial parts of the plant have begun to wither and die. Dig up the roots of annuals once their growth period is over and the roots of perennials in their second and third years, when all the active substances should have developed.

Dig the soil away from the root, avoiding cutting or bruising it. Cut off the parts or the amount needed and cover up the other parts of the root again. If the root is very large and not all of it is needed, part of it can usually be harvested without pulling up the entire plant.

Wash off all the soil and dry the root, so that it can dry as soon as possible. Large, thick roots should be cut into 2.5 cm thick slices before being dried, both to reduce the drying time and because it is easier to cut the roots before they dry.

BARK

Bark is most potent in the spring and autumn when its moisture content is greatest. Bark is best harvested on a damp day, when it peels easily off the trunk and branches. It is best to take the bark off thick branches of robust trees, or from the trunks of trees which have been felled. Since it kills the tree if the bark is taken in a ring surrounding the entire trunk, avoid peeling the bark from the trunk of a living tree. Clean the bark well and cut it into small pieces to dry.

DRYING

Plants must be dried as quickly as possible after gathering, because chemical changes begin as soon as the leaves or flowers are removed from the plant. Enzymes which previously contributed to creating active substances can now begin to break down those same substances.

Plants must not be dried too quickly or too slowly, because some substances, especially volatile oils are lost at too high a temperature.

The drying time varies for each plant; some dry in four days while others need up to three weeks to dry fully.

LEAVES AND FLOWERS

Any dirt on leaves or flowers should be wiped off with a dry cloth rather than rinsed off with water.

Dry the herbs in a warm, dark place where the air can circulate around them. The temperature should be around 32°C for the first two days and 25°C after that. It is best to spread flowers and leaves out on a net or other porous material where air can circulate freely around them. Turn the plants regularly so that they will dry evenly.

When the entire aerial part of the plant is gathered, it is best hung up to dry.

ROOTS AND BARK

Wash and dry the roots and bark well before spreading them out. Both roots and bark should be cut into small pieces to accelerate drying and facilitate the preparation of decoctions and tinctures later.

Roots and bark need a high temperature of as much as 50°C to dry well, and should be dried all together in a net or similar material so the air can circulate evenly around all of the parts.

SEEDS AND FRUIT

Seeds are dried the same way as leaves and flowers. They dry quickly in a warm, dark place where the air can circulate around them. Generally seeds dry within two weeks.

Berries and fruits take a longer time to dry and need to be turned often to speed up drying.

STORAGE OF DRIED HERBS

The herbs are best stored in dark, sealed glass containers or glazed crockery containers. Place the herbs in a container as soon as they are well dried, label the container clearly with the contents and date. Store the containers in a dark, cool place. Herbs will keep for two to three years if handled properly.

They are most potent, however, the first year after drying, so that it is recommended to renew them annually. This applies especially in the case of the aerial parts of plants. Roots and bark keep better and therefore only need be replaced every two to three years.

MIXING OF HERBAL MEDICINES

Herbal pharmacology is a study which can only be described to a very limited extent in this book. It is very likely that when humans began using herbs for health purposes they simply consumed them as they occurred in nature. There was a gradual increase in human knowledge of the usefulness of individual plants and plant parts for medicinal purposes, and how they could be prepared to be most effective. The following are some of the most common methods of preparing herbal medicines.

INFUSIONS

To make an infusion, a heat-resistant container with a tight-fitting lid should be used. Do not use aluminium pots or unglazed crockery.

To make one cup of infusion, place 1-2 tsp of the herb in a cup of water. To prepare an infusion for three days, use 100 g of herb per litre of water, i.e. a proportion of 1:10.

Infusions do not keep for long and therefore it is advisable not to prepare more than a three-day dosage at a time.

PREPARING AN INFUSION

- *Place the herbs in a container.*
- *Pour boiling water over them.*
- *Close the container and allow it to stand for 20-30 minutes to seep.*
- *Filter through cheesecloth and squeeze the herbs well to extract as much as possible of their essence.*
- *Store the infusion in the refrigerator if more than a single day's dose has been prepared.*
- *An adult dosage is 100 ml three times daily, unless otherwise indicated.*

Infusions are generally made from the aerial parts of the plants, i.e. the leaves flowers and stems. In these parts of the plants the active substances are readily soluble.

If infusions are to be made from seeds, roots or bark to utilise substances which cannot withstand boiling (generally decoctions are made from these plant parts) the parts must be crushed well, or even ground to a powder before infusing. To make an infusion from fresh herbs, which have a high water content, use three times the quantity of the dry herbs recommended.

DECOCTION

Decoction is the method used to extract the essences from tougher plant parts, e.g. the roots, berries and seeds. Decoctions are used in the same manner as infusions.

Generally around one-third of the water evaporates in boiling, so that to prepare a decoction for three days requires 1.3 litres of water for every 100 g of dried herbs or 300 g of fresh herbs. An adult dosage is 100 ml three times daily, unless otherwise indicated.

PREPARING A DECOCTION

- *Place the herbs in a pot (never use aluminium pots).*
- *Pour cold water over them.*
- *Place a lid on the pot and bring the contents slowly to a boil.*
- *Simmer for 20-60 minutes.*
- *Remove from heat and allow the pot to stand for an additional 10 minutes with the lid on.*
- *Filter through cheesecloth and squeeze the herbs well to extract as much as possible of their essence.*
- *Store the decoction in a closed container in the refrigerator if more than a single day's decoction is prepared at a time.*

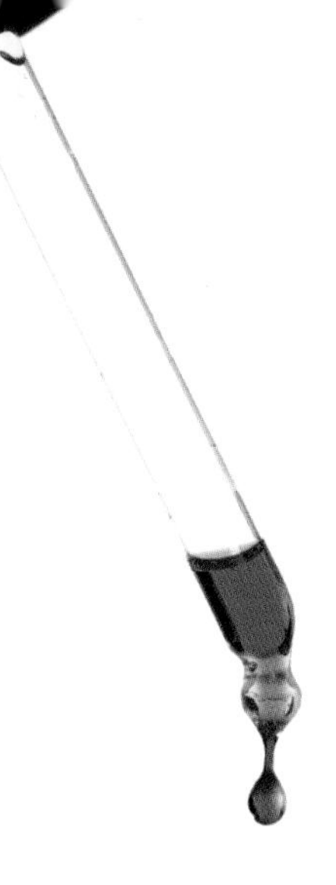

TINCTURE

A tincture is an herbal medicine in alcohol (or vinegar) solution. Some substances in herbs dissolve more readily in alcohol than in water. Tinctures can be prepared in various ways and the rules on proportions and strength of the alcohol vary depending upon the plant. It is advisable to prepare the tincture always in the same manner, unless a recipe instructs otherwise. Alcohol with 45% spirit content is sufficiently strong to dissolve most active substances in plants and also increases the storage life of the medicinal mixture. A tincture of this strength can be preserved for many years.

Tinctures have an advantage over infusions in that the plants are utilised better, the dosage can therefore be smaller and the medicine can be stored longer.

A tincture can be used in a variety of ways. It can be taken orally as is or can be diluted with hot or cold water. A tincture can be added to bathwater or mixed with ointments and creams for external use. Tinctures are generally prepared in a proportion of 1:5, i.e. 100 g of herbs to 500 ml of alcohol.

A tincture can also be prepared by using wine, especially white wine, instead of strong spirits. The tincture is prepared in the same manner but the taste can be considerably better and in some instances more suitable, especially for persons with a sensitive stomach. A tincture made from wine, however, has a shorter shelf life.

PREPARING A TINCTURE

- *Place 100 g of finely chopped or ground (dried) herbs in a container with a tight-fitting lid. Double the amount if the herbs are fresh.*
- *Pour 500 ml of alcohol over the plants. Often the herbs need to be pressed down so that they are covered entirely by the alcohol.*
- *Close the container securely, label and date it.*
- *Store the container in a warm location for 2-3 weeks, stirring the contents gently once a day.*
- *At the end of that time, filter the liquid through cheesecloth and squeeze the herbs well to extract as much as liquid as possible. It is recommended that the cheesecloth be placed over a funnel to filter the liquid. A good juice press can be used to extract all the juice from the herbs.*
- *Store the tincture in dark, tightly sealed bottles.*

TINCTURE MADE FROM VINEGAR

Cider vinegar is recommended for a vinegar tincture. Cider vinegar itself has health promoting characteristics and therefore a tincture made from this vinegar can be useful in many instances, especially for all types of rheumatism and cancer. A vinegar tincture is prepared in the same manner as other tinctures, with the exception that the herbs are allowed to steep in the vinegar for longer, up to six weeks, before the liquid is filtered off. The proportions and dosages are the same as in a conventional tincture.

SYRUP

Syrup can often be given to children who are reluctant to take other herbal medicines. The syrup should preferably be made from honey, which has the advantage over sugar of being rich in vitamins and antibacterial. Of the many types of honey, thick liquid or solid honey is best.

PREPARING A SYRUP

- *Prepare 1/2 litre of an infusion with double strength, i.e. 1:5 (see p. 103). Filter and cool.*
- *Pour the infusion into a pot and add 125 g of honey.*
- *Allow this to thicken slowly over low heat and stir gently until the mixture becomes syrupy.*
- *Now and again skim off any froth which forms.*
- *Store the syrup in a closed glass jar and use a dose of 1 tsp as needed.*

PREPARING HERBAL MEDICINES FOR EXTERNAL USE

Medicines for external (topical) use contain active substances which are absorbed through the skin or mucous membrane, e.g. in the vagina or rectum. There are various methods and recipes for herbal medications to use externally.

BATHS

Since ancient times, baths have been used extensively for herbal healing. They are especially suitable for infants and persons who have difficulty taking herbal medicines orally.

Pure herbal oils are often used for baths. Some of them have an irritant effect and a few drops of pure herbal oil in the bath is sufficient. Allow the oil to mix well with the bathwater before getting into the bath.

BATHS CAN BE PREPARED IN TWO WAYS:

1. *Prepare a strong infusion (1:5, or a decoction, depending upon what parts of the plants are used) and then add it to the bathwater. One litre of infusion is needed for an average bathtub.*
2. *Wrap the herbs which are to be used in cotton gauze and allow hot tap water to flow through the herbs in the gauze into the bath. Cool the bath afterwards as necessary, running the cold water through the gauze as well. Use 150-200 g of herbs in a full bath. This method can only be used when a bath is prepared with leaves, flowers and small stems of plants, because roots and harder parts of the plants need to be boiled to extract the active substances from them.*

EYEWASH

Eyewashes are either an infusion or decoction. If an infusion is used, it should be prepared fresh each time. The infusion must be weak, use ½ tsp of herbs in ½ cup of water. Be sure to filter the infusion very well and cool it before using. Place the infusion in two containers and use one for each eye. Always rinse both eyes in the case of infection.

Bear in mind that the eyes are very sensitive organs, so always consult a physician for a proper diagnosis before trying herbal remedies.

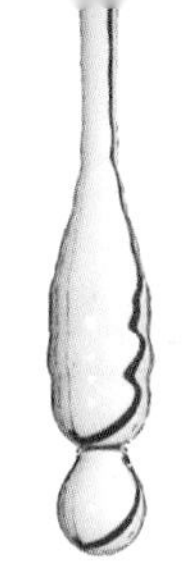

VAGINAL DOUCHE

A vaginal douche is used to treat vaginal infections, e.g. mycosis (yeast infections).

Use an undiluted infusion or diluted tincture. The infusion must be prepared fresh each day. For application, it is easiest to use special douche syringes. Use 100 ml of infusion twice or three times daily, at body temperature.

A tincture is simpler to use because it can be stored ready to use when needed. Mix 5 ml of tincture with 100 ml of warm water and douche twice or three times daily.

If douching does not remedy the condition within a week, seek medical advice. If there is reason to suspect a sexually transmitted disease, the advice of a physician should be sought before trying herbal remedies.

SUPPOSITORIES

Suppositories are used, for instance, to soften and heal mucous membrane in the rectum. For this purpose suppositories are used containing vulnerary plants, e.g. chickweed or coltsfoot. Suppositories are also often used for haemorrhoids, and in this case astringent herbs are used, e.g. alpine lady's mantle and alpine bistort, as well as anti-inflammatory herbs, such as willow. Suppositories can also be used for vaginal inflammation and pain. It is simplest to make suppositories from finely ground powder prepared from the herbs to be used mixed with a suitable base material to bind the powder. The best base material is cocoa butter which is easy to shape and also has the useful quality of melting at body temperature.

HOME-MADE SUPPOSITORIES

- *Warm the cocoa butter so the herbal powder can easily be mixed with it.*
- *Pour the mixture into suppository moulds, which can be made from aluminium foil; or cool it so that suppositories can be formed by hand.*
- *Store the suppositories in their moulds or in plastic in a refrigerator.*

OINTMENT

Ointments are a semi-liquid mixture of a medicine and fatty substance for external use. Most herbs can be used in ointments. To combat itchy rash, fresh chickweed, stinging nettle and northern dock are useful, and yarrow for burns. As demulcent, coltsfoot is recommended; red clover is useful to treat skin cancer; and dandelion and common sundew are used for warts and corns.

The simplest way to prepare an ointment is to use Vaseline as a base. A mixture of cocoa butter and melted beeswax is also good for this purpose.

If ointment is to be used in quantity, it may be advisable to purchase ready-made ointment with which an infusion or tincture can be mixed. The ointment must be made of natural substances and thick enough so that liquid can be added without making it separate.

VASELINE OINTMENT

- *Melt 200 g of Vaseline over low heat.*
- *Add 60 g of herbs and bring to a boil.*
- *Boil gently for 10-15 minutes, stirring gently for the whole time.*
- *Filter through cheesecloth and press all the juice from the herbs.*
- *Pour the mixture into a jar with a tight lid, cool it and place waxed paper over the top of the jar before putting the lid on so it will keep better. Store in a refrigerator.*

TAR SALVE

Tar salve can be prepared from various herbs and is very potent. Red clover salve has proven the most effective, and is used to treat various types of skin cancers. ***The tar is then applied as necessary.***

LOTION

Lotions are useful for tense and arthritic muscles and joints. Because of the lotion's characteristics, the skin can easily absorb those substances which relieve and heal. ***Shake the bottle well before using the lotion because it tends to separate.***

EXAMPLE OF A LOTION

- *Pour 50 ml of almond oil in a bowl.*
- *Add 35 ml of herbal tincture.*
- *Mix well and pour into a coloured glass bottle.*

RED CLOVER SALVE FOR SKIN CANCERS

- *Place 200 g of dried red clover flowers in a slow-cooker (crock pot) with 2 litres of water.*
- *Simmer on low heat for 48 hours.*
- *Press all the liquid from the pulp and return it to the pot on high heat.*
- *Allow this to boil until only a black, tarry residue remains on the bottom of the pot. This can take up to a day and a half.*

OILS

Oils can be prepared in two ways. Pure essential oils have to be processed from herbs by a complex process of distillation, which is not recommended as general practice. Such oils are best purchased as needed. However, herbal oils can be prepared more simply, although they will not be as rich in active substances as when prepared by distillation.

Oils can be used in the same ways as ointments on arthritic joints and tense muscles and on all rashes and sensitive skin. Oils can be prepared from all herbs recommended for external use.

OILS PREPARED FROM HERBS

- *Chop the herb to be used to prepare the oil and place it in a clear glass jar.*
- *Pour almond, olive or sunflower oil over the herbs until the volume of the oil is about twice that of the herbs.*
- *Close the jar and store in a warm place, preferably in sunlight, for 2-3 weeks. The bottom of the jar should preferably be embedded in sand, which will reduce the heat fluctuations.*
- *Stir the contents of the jar daily.*
- *Filter the oil from the jar and press the pulp well to extract all the oil.*
- *Pour the herbal oil into dark bottles, label and date them and store in a cool place.*

POULTICES

Poultices are placed on arthritic joints and wounds which are slow to heal. Both fresh and dried herbs can be used for poultices. When fresh herbs are used, the plant parts are crushed well and placed directly on the skin, then covered with a hot, moist gauze to fix them in place. When dried herbs are used, a paste must be prepared from them before they are placed on the skin. To make the paste, the herbs are ground to a powder, then a small amount of boiling water is poured over them and they are left for five minutes to absorb the water. Next the herbs are mixed thoroughly with very thick oatmeal porridge. Spread the warm poultice on the tender area, cover with gauze and keep the poultice warm by covering it with plastic and a heating pad.

COMPRESSES (DRESSINGS)

A compress is used in a manner similar to a poultice, but the herbs in compresses do not come into direct contact with the skin. Wrap the herbs to be used in a clean cotton cloth or gauze (cheesecloth). Soak the compress in boiling water for 5-10 minutes, wring it slightly and place it as hot as possible on the sore area. The compress must be hot and must be changed when it cools, or it can be covered with plastic and kept hot with a heating pad.

Herbs which are suitable for compresses are vulneraries and herbs that stimulate circulation.

DOSAGES AND PREPARATION OF MEDICINES

DOSAGE

The dosage is indicated for each herb in the main section of the book. Here Iceland moss is taken as an example, with explanations provided for each item.

DOSAGES:
TINCTURE: ***1:5****, 25% alcohol*
2–5 ml *three times daily.*
INFUSION OR DECOCTION:
1:10, ***100 ml*** *three times daily or* ***1 tbsp*** *in 1 cup of water, drunk three times daily. The taste of Iceland moss is less bitter if it is boiled in milk, but boiling in water is better for the lungs.*
Poultices *for external use.*

- *One part herbs (by weight) to 5 parts 25% alcohol spirits (e.g. 100 g herbs to 500 ml 25% alcohol).*
- *Of this tincture, 2-5 ml are taken three times daily.*
- *Infusion: One part herbs to 10 parts boiling water, or one tbsp to one cup of water.*
- *Decoction: See Decoction in the section on mixing herbal medicines.*
- *Dosage: 100 ml, drunk three times daily*
- *See the section on mixing herbal medicines on preparing poultices.*

HERBAL MEDICINE MIXTURE

An herbal medicine mixture is either a mixture of an infusion and decoction, or a tincture mixture (see the following pages).

First, the best way to prepare a mixed herbal tincture (i.e. from several different herbs) is discussed, then how to prepare a mixture of herbs used for either an infusion or a decoction.

MIXED HERBAL TINCTURE

The best way to prepare a mixed herbal tincture is to first prepare a tincture of each individual herb and then mix them in the proper proportions.

MIXTURE OF INFUSIONS AND DECOCTIONS

- *First, prepare the herbs which need to be boiled:*
- *Place 50 grams of the herbs in 500 ml of water in a pot (1:10) and boil for 20 minutes.*
- *Then take the herbs which should not be boiled:*
- *Pour 300-500 ml of boiling water over 50 grams of the chosen herbs, close the container and allow to stand for 20 minutes. Mix the decoction and the infusion together and press out the juice. Store the mixture in the refrigerator and drink 100 ml three times daily.*

AN EXAMPLE OF A HERBAL MIXTURE FOR INSOMNIA:

50 ml valerian (tincture) page. 84

50 ml yarrow (tincture) page. 92

This makes a total of 100 ml which is then stored and taken in doses of 5 ml twice daily.

LISTS AND DEFINITIONS

DEFINITIONS

acids: substances which release hydrogen ions in a water solution. Weak acids, generally in the form of salts or esters, are very common in the plant kingdom. Acids which are useful for their medicinal qualities include isovaleric acid in valerian, which is used for its sedative effects, and formic acid in stinging nettle, which is used for its irritant effect on the skin. Acids with a strong scent are also common and the plants containing them are used, among other things, for respiratory ailments.

alkaloids: nitrogenous basic organic compounds, most of which are pharmacologically active or toxic. Many plants contain alkaloids, but the role of most of the substance in the plants themselves is uncertain. Each alkaloid has its own special effect, making it impossible to generalise as to their impact. Some alkaloids are very well known and strong drugs, such as morphine, quinine, nicotine, ergotamine, ephedrine and cocaine.

amino acids: various organic compounds which are the basic units in protein.

anthraquinones, anthraquinone glycosides: a class of phenol compounds. Plants containing anthraquinones have long been used for both dyeing and as laxatives. Plants containing anthraquinones are useful to treat chronic constipation resulting from a lazy colon. Anthraquinones irritate the colon wall and strengthen its movements.

antispasmodic: a substance which counteracts muscle cramps or seizures.

astringent substances, astringents: substances which cause a contraction in body tissues, thereby stopping or significantly reducing the flow of blood or other fluids. An astringent skin cream tightens the skin and reinforces it, causing small skin lesions to close more rapidly.

birchwater/-syrup: the sap which seeps from the birchwood capillaries if the bark is scored.

bitter principles: a collective term for a variety of substances which all have a bitter taste. Bitter principles stimulate the taste buds which are sensitive to bitter or acrid taste and stimulate secretion of saliva and digestive fluids. Plants which contain bitter principles therefore stimulate appetite, flow of gall and dilution of gall. They have a positive effect on the pancreas and regulate the production of the pancreatic hormones insulin and glucagon. Many bitter herbs are highly strengthening and were often the main ingredient in elixirs.

carotene: yellow or reddish colouring substance in many plants, e.g. carrots; a precursor of vitamin A in the body.

choline: one of the B vitamins; necessary for liver function.

coumarin: a substance derived from benzopyrene. Coumarin is antibacterial and its derivative, dicoumarol, is an anticoagulant.

cyanoglycosides, cyanogenic glycosides: a group of glycosides which release cyanide in hydrolysis and can be fatal in large doses. In small quantities they have useful effect, for instance, antispasmodic and sedative. They slow the heartbeat and lower blood pressure, have a positive effect on digestion and are mildly analgesic.

diuretic substance: substances in various plants which increase the amount of urine in the body. The diuretic effects of plants vary, some of them stimulate the flow of blood to the kidneys and therefore production of urine increases, while other stimulate the entire circulatory system. Still other plants contain substances which the kidneys excrete, which requires additional fluid ending up in the urine. There are also plants which affect the kidneys directly thereby promoting increased urine production. Highly effective diuretic plants include dandelion (the leaves), couchgrass, birch, juniper and heather.

elixir: a drink made formerly from various stimulating herbs which was intended to cure all illnesses and even ensure longevity.

enzyme: a substance produced by a living organism that acts as a catalyst for biochemical reactions, including metabolizing compounds into usable forms.

flavonoids: a varied group of substances from the phenol class, the most common phenols in the plant kingdom. Flavonoids are important for medical purposes. Some are antispasmodic, others diuretic and still other affect the heart and respiratory system. Isoflavonoids have a very similar structure to steroids, which explains the effect of many of them on the body's hormone system. Bioflavonoids (vitamin P), in particular hesperidin and rutin, have been studied extensively because of their strengthening effect on blood vessels and capillaries. It has also been discovered that vitamin C does not occur naturally without bioflavonoids, providing support for the theory subscribed to by many persons that vitamin C alone does not have a completely satisfactory affect on the body.

gluten: a protein composite found in the endosperm of various grains. Gluten can cause an allergic reaction in the small intestine, impeding the absorption of nutrients. Gluten intolerance is most common in infants when they begin to eat grains.

glycosides: organic chemical compounds in which a sugar is bound to a non-carbohydrate moiety. Some glycosides are important medicines, but many are toxic.

herb: (1) a plant with soft parts, in contrast to woody plants; (2) a medicinal plant of any type, both woody and non-woody.

inulin: a polysaccharide which the body does not digest but is excreted from the body through the kidneys.

isovaleric acid: → valeric acid.

mannitol: a sugar alcohol which is used, for instance, as a carrier in pharmaceutical tablets. Mannitol is not digested but is excreted from the body through the kidneys, drawing out liquid as it does so. As a result, plants containing mannitol, e.g. couchgrass, are used as diuretic herbs.

mucilages: complex saccharides which produce a gluey substance mixed with water. Some of them are important medicines and all of them have a local effect on the body as they are decomposed in the digestive tract. Mucilages are demulcent and protect mucous membranes, reducing irritation and promoting healing.

oestrogen: a collective name for female hormones produced in the ovarian follicles and elsewhere; they promote the development of secondary female characteristics, regulate menstruation and prepare the sexual organs for the development of an embryo.

pectin: a polysaccharide, used in particular as a gelling agent. Pectin is not digested by the body but remains to boost stool volume and is therefore used as a laxative. It has been revealed to reduce absorption of cholesterol from food and has a local antibacterial effect.

phenols and phenolglycosides: substances derived from carbolic acid. Phenols are generally antibacterial, cause blisters, are anaesthetic and cause protein to gel. Salicylic acid, which is found either free or in compounds in many plants (e.g. salisin in willows), is a phenol, as is thymol in wild thyme, and is an antiseptic and local anaesthetic.

plant resins: → resins.

resins, plant resins: the exudate of various plants which often becomes hard and brittle when it dries, but softens with heating. Resins are not soluble in water and therefore are more useful in an alcohol solution. Resins are considered to stimulate the activities of white blood cells.

saccharides (sugars): a varied group of organic compounds of carbon, hydrogen and oxygen. Saccharides are the body's main source of energy, and affect many aspects of its functioning. Those saccharides which are particularly useful medically are resins and mucilage. Plant saccharides have a demulcent and coating effect on the body tissue they come into contact with. Research on the effect of saccharides on the body has shown that they also affect tissue other than those they directly contact through their complex, involuntary responses. The effect of saccharides on the body is therefore not only limited to the skin and digestive tract, they also have a demulcent and as a result relaxing effect e.g. on the respiratory and urinary organs.

salicylate: a salt or ester of salicylic acid.

salicylic acid: a substance found especially in willow species which is the active substance in various analgesic medicines, including aspirin. In addition it is used as a preservative, to treat various rheumatic conditions and it appears demonstrated that it counteracts arterial plaque build-up and thrombosis.

saponins: complex glycosides which have been used to manufacture a variety of soaps. Saponins have also been used in pharmaceutical production because they include many precursors of hormones. Saponins include a steroidal subset, saraponins and triterpenoid saponins. Many saponins are expectorants; they stimulate the absorption of nutrients from food and many plants which contain saponins are thought to strengthen various glands of the body and their functioning.

silicic acid: a weak acid of hydrogen, silicon and oxygen which is found, for instance, in soils. It occurs in the cells of various plants, especially of the grass, horsetail and forget-me-not families. Silicic acid is used to correct deficiency of the same, e.g. resulting from poor nutrition. Silicic acid is found in hair, nails and connective tissue.

sorbitol: a sugar alcohol that is found especially in plants of the Rose family. It is used as a sweetener (for diabetics) and in the production of vitamin C.

steroids: organic lipid soluble compounds based on a core of 17 carbon atoms bonded together to form four fused rings. They are common as active substances in plants and animals. Steroids include, for instance, sex hormones, corticosteroids, bile salts, cholesterol and vitamin D.

tannins: substances which have an astringent effect on the body, i.e. they cause protein to condense or gel, as in tanned leather, and cause living tissue to contract. Tannins have a restrictive effect on most alkaloids and are therefore used as an antidote to toxicity caused by alkaloids. In large doses, tannins can have a very irritating effect on mucous membrane, but in small doses they cause mucous membrane to contract and make it therefore less accessible to irritants while at the same time reducing secretions from it. Plants which contain tannins are therefore used to treat inflammation and pain, both internally and externally, e.g. for gastritis and ulcers which fail to heal due to digestive fluids which are constantly in contact with the mucous membrane or other irritation. Tannins are used for persistent diarrhoea to reduce irritation in the mucous membrane of the intestines and also to form a protective film over minor burns.

valeric acid: a saturated organic acid found in some plants, including valerian. The acid occurs both free and as an ester in volatile oils and is used, for instance, in medicines (e.g. the sedative valeriana) and as a flavour. Isovaleric acid is one form of valeric acid.

volatile oils: volatile substances which develop in plants; a very complex mixture of oxygenated carbohydrates and their polymers, often with terpene as their basic unit. The scent of plants comes from their volatile oils, and the oils themselves have varying effects on the body. Some oils are antibacterial, while others have a strong effect on various body systems. They can stimulate blood flow and have a wide-reaching effect, for instance, on the nervous system, digestive organs, urinary organs and lungs. Most of the oils stimulate the formation of white blood cells.

PROTECTED PLANTS IN ICELAND

Common tormentil, Icel. **blóðmura**	*Potentilla erecta*
Wolf's foot clubmoss, Icel. **burstajafni**	*Lycopodium clavatum*
Greenland primrose, Icel. **davíðslykill**	*Primula egaliksensis*
Glossy moonwort, Icel. **dvergtungljurt**	*Botrychium simplex*
Common twayblade, Icel. **eggtvíblaðka**	*Listera ovata*
Herb Paris, Icel. **ferlaufungur**	*Paris quadrifolia*
Saltmarsh rush, Icel. **fitjasef**	*Juncus gerardi*
Lesser sea spurrey, Icel. **flæðarbúi**	*Spergularia salina*
Glaucous dog rose, Icel. **glitrós**	*Rosa dumalis*
Hudson Bay sedge, Icel. **heiðastör**	*Carex heleonastes*
Parsley fern, Icel. **hlíðaburkni**	*Cryptogramma crispa*
Foliolose saxifrage, Icel. **hreistursteinbrjótur**	*Saxifraga foliolosa*
Yeo, Icel. **hveraaugnfró**	*Euphrasia calida*
Green spleenwort, Icel. **klettaburkni**	*Asplenium viride*
Common heath grass, Icel. **knjápuntur**	*Sieglingia decumbens*
Northern stichwort, Icel. **línarfi**	*Stellaria calycantha*
Pyramidal bugle, Icel. **lyngbúi**	*Ajuga pyramidalis*
Arctic poppy, Icel. **melasól** (with white and pink flowers)	*Papaver radicatum ssp. Stefanssonii*
Wilsons filmy fern, Icel. **mosaburkni**	*Hymenophyllum wilsonii*
Marsh bedstraw, Icel. **mýramaðra**	*Galium palustre*
Forked spleenwort, Icel. **skeggburkni**	*Asplenium septentrionale*
Hard fern, Icel. **skollakambur** (geothermal variety)	*Blechnum spicant var. fallax*
Common dog-violet, Icel. **skógfjóla**	*Viola riviniana*
Wood sorrel, Icel. **súrsmæra**	*Oxalis acetosella*
Maidenhair spleenwort, Icel. **svartburkni**	*Asplenium trichomanes*
Amphibious bistort, Icel. **tjarnablaðka**	*Persicaria amphibia*
Peduncled water-starwort, Icel. **tjarnabrúða**	*Callitrice brutia*
Large yellow-sedge, Icel. **trjónustör**	*Carex flava*
Pygmyweed, Icel. **vatnsögn**	*Crassula aquatica*
Field garlic, Icel. **villilaukur**	*Allium oleraceum*

In Iceland the Nature Conservation Council is authorised by law to declare individual plant species as protected. Protection may apply to a specific location or the entire country. Currently 31 plant species are protected. It is prohibited to remove shoots, leaves, flowers or roots of these plants, to crush them, dig them up or disturb them by any means.

BIBLIOGRAPHY

Ágúst H. Bjarnason: *Íslensk flóra með litmyndum.* Iðunn, 1983.

Bensky, Dan and Andrew Gamble: *Chinese Herbal Medicine: Materia Medica.* Eastland Press, 1993.

Björn Halldórsson: *Gras-nytjar*, Kaupmannahöfn, 1783.

Björn L. Jónsson: *Íslenskar lækninga- og drykkjarjurtir.* Náttúrulækningafélag Íslands, 1977.

British Herbal Pharmacopoeia. British Herbal Medicine Association, 1971.

Campion, Kitty: *Handbook of Herbal Health.* Sphere Books Limited, 1985.

DK Pocket Encyclopedia. Herbs. Dorling Kindersley, 1990.

Fluck, Hans: *Medicinal Plants.* W. Foulsham & Co Ltd., 1976.

Grieve, M.: *A Modern Herbal.* Dovec Publications, Penguin Books, 1980.

Hudson, Paul: *Mastering Herbalism.* Stein & Day Publishers, 1974.

Hörður Kristinsson: *Plöntuhandbókin: blómplöntur og byrkningar.* Örn og Örlygur, 1986.

Jónas Jónasson frá Hrafnagili: *Íslenzkir þjóðhættir.* Ísafoldarprentsmiðja, 1961.

Launert, Edmund: *The Hamlyn Guide to Edible & Medicinal Plants of Britain and Northern Europe.* Hamlyn, 1981.

Lítil ritgjörð um nytsemi nokkurra íslenzkra jurta eptir ýmsa höfunda. Safnað hefur Jón Jónsson garðyrkjumaður. Einar Þórðarson, 1880.

Lust, John: *The Herb Book.* Bantana Books, 1974.

Lúðvík Kristjánsson: *Íslenzkir sjávarhættir* I–V. Bókaútgáfa Menningarsjóðs, 1980–1986.

Mabey, Richard: *New Herbal.* ElmTree Books, 1988.

Macdonald Encyclopedia of Medicinal Plants. Macdonald & Co Ltd., 1984.

Mills, Simon: *Dictionary of Modern Herbalism.* Thorsons Publishers, 1985.
Naturopathic Medical Series Vol I–III. Electrical Medical Publications, 1985.
Potters New Cyclopaedia of Botanical Drugs. Health Science Press, 1907.
Sveinn Pálsson: *Ferðabók Sveins Pálssonar: Dagbækur og ritgerðir 1791–1797.* Örn og Örlygur, 1983.
Weizz, RF.: *Herbal Medicine.* Beaconsfield Publishers Ltd., 1985.
Yeung, Him-Che: *Handbook Of Chinese Herbs And Formulas.* Institute of Chinese Medicine, Ca. U.S.A., 1985.